Introduction to High Temperature Oxidation of Metals

N Birks and G H Meier
Professors of Metallurgical Engineering
University of Pittsburgh

Edward Arnold

First published in 1983
by Edward Arnold (Publishers) Ltd
41 Bedford Square, London WC1B 3DQ

British Library Cataloguing in Publication Data
Birks, N.
Introduction to high temperature oxidation of metals.
1. Oxidation 2. Metallic oxides
3. Metals at high temperature
I. Title II. Meier, G. H.
660.2′9′687 QD501

ISBN 0-7131-3464-X

Printed in Great Britain by
Thomson Litho Ltd, East Kilbride, Scotland

Preface

Few metals, particularly those in common technological applications, are stable when exposed to the atmosphere at both high and low temperatures. Consequently, most metals in service today are subject to deterioration, either by corrosion at room temperature or by oxidation at high temperature. The degree of corrosion varies widely. Some metals, such as iron, will rust and oxidise very rapidly whereas other metals, such as nickel and chromium, are attacked relatively slowly. It will be seen that the nature of the surface layers produced on the metal plays a major role in the behaviour of these materials in aggressive atmospheres.

The subject of high temperature oxidation of metals is capable of extensive investigation and theoretical treatment. It is, therefore, normally found to be a very satisfying subject to study. The theoretical treatment covers a wide range of metallurgical, chemical, and physical principles and can be approached by people of a wide range of disciplines who, therefore, complement each other's efforts.

Initially, the subject was studied with the broad aim of preventing the deterioration of metals in service, i.e. as a result of exposing the metal to high temperatures and oxidising atmospheres. In recent years, a wealth of mechanistic data has become available. These data cover a broad range of phenomena, e.g. mass transport through oxide scales, evaporation of oxide or metallic species, the role of mechanical stress in oxidation, growth of scales in complex environments containing more than one oxidant, and the important relationships between alloy composition and microstructure and oxidation. Such information is obtained by applying virtually every physical and chemical investigative technique to the subject.

In this book the intention is to introduce the subject of high temperature oxidation of metals to students and to professional engineers whose work demands familiarity with the subject. The emphasis of the book is placed firmly on supplying an understanding of the basic, or fundamental, processes involved in oxidation.

In order to keep to this objective, there has been no attempt to provide an exhaustive, or even extensive, review of the literature. In our opinion this would increase the factual content without necessarily improving the understanding of the subject and would, therefore, increase both the size and price of the book without enhancing its objective as an introduction to the subject.

Extensive literature quotation is already available in books previously published on the subject and in review articles. Similarly the treatment of techniques of investigation has been restricted to a level that is sufficient for the reader to understand how the subject is studied without involving experimental details. Such details are available elsewhere as indicated.

After dealing with the classical situations involving the straightforward oxidation of metals and alloys in the first five chapters, the final chapters extend the discussion to reactions in mixed environments, i.e. containing more than one oxidant, and to reactions involving a condensed phase as in hot corrosion. Finally, examples of the application of the principles described are given under the headings of atmosphere control and the scaling and decarburisation of steels, both of which are important commercial subjects.

Pittsburgh
1982

NB
GHM

Contents

1
Methods of investigation

Introduction

The purpose of oxidation experiments is generally to assess the reaction kinetics and the mechanism of oxidation of a metal or alloy under a particular set of exposure variables, e.g. temperature, pressure, gas composition, etc. The simplest approach is to expose a specimen of known mass and dimensions in a furnace for the appropriate time, remove the specimen, and allow it to cool. The specimen may then be weighed to assess the extent of oxidation and the oxidation morphology may be observed by various X-ray and metallographic techniques.

Although this procedure is simple, one drawback is that the start time for the reaction cannot be established accurately. Several starting procedures are commonly used.

(a) The sample may simply be placed in the heated chamber containing the reactive atmosphere.
(b) The sample may be placed in the cold chamber containing the reactive atmosphere and then heated.
(c) The sample may be placed in the cold chamber which is then evacuated, or flushed with inert gas, heated, and then, at temperature, the reactive gas is admitted.

In all cases, the start of the reaction is in doubt either due to the time required to heat the specimen or due to the inevitable formation of thin oxide layers, even in inert gases or under vacuum, especially in the case of the more reactive metals, so that when the reaction is started by admitting the reactive gas an oxide layer already exists.

Attempts have been made to overcome this by heating initially in hydrogen which is then flushed out by the reactive gas. This also takes a finite time and thus introduces uncertainty concerning the start of the reaction.

Thin specimens may be used to minimise the time required to heat the specimen. In this case, care should be taken that they are not so thin and, therefore, of such low thermal mass that the heat of reaction, released rapidly during the initial oxidation period, causes severe specimen overheating.

The uncertainty concerning the start of the reactions usually only affects results for short exposure times, of up to about ten minutes, and becomes less noticeable at longer times. However, in some cases, such as selective oxidation of one element from an alloy, these effects can be quite long-lasting. In practice, specimens and procedures must be designed with these factors in mind.

Many early investigations were simply concerned with oxidation rates and not with oxidation mechanisms.

The rate of formation of an oxide on a metal according to the reaction

$$2M + O_2 = 2MO$$

may be investigated by several methods. The extent of the reaction may be measured by

(a) the amount of metal consumed,
(b) the amount of oxygen consumed,
(c) the amount of oxide produced.

Of these only (b) can be assessed continuously and directly.

(a) The amount of metal consumed In practice this may be assessed by observing either the mass loss of the sample or the residual metal thickness. In both cases, the sample must be removed from the furnace thus interrupting the process.

(b) The amount of oxygen consumed This may be assessed by observing either the mass gain or the amount of oxygen used. Both of these methods may be used on a continuous and automatic recording basis.

(c) The amount of oxide produced This may be assessed by observing the mass of oxide formed or by measuring the oxide thickness. Of course, in this case, it is necessary to destroy the sample, as it is with method (a).

Of the above methods, only those involving measurement of mass gain and oxygen used give the possibility of obtaining continuous results. All the other methods require destruction of the specimen before the measurement can be achieved and this has the drawback that, in order to obtain a set of kinetic data, it is necessary to use several samples. Where the sample and the method of investigation are such that continuous results can be obtained, one sample will give the complete kinetic record of the reaction.

When representing oxidation kinetics, any of the variables mentioned above can be used, and can be measured as a function of time because, of course, they all result in an assessment of the extent of reaction. Nowadays, it is most general to measure the change in mass of a specimen exposed to oxidising conditions.

It is found experimentally that several rate laws can be identified. The principal laws are

(a) Linear law
(b) Parabolic law
(c) Logarithmic law

The linear law, for which the rate of reaction is independent of time, is found to refer predominantly to reactions whose rate is controlled by a surface reaction step or by diffusion through the gas phase.

The parabolic law, for which the rate is inversely proportional to the square root of time, is found to be obeyed when diffusion through the scale is the rate determining process.

The logarithmic law refers to the formation of very thin films of oxide only, that is between 20 and 40 ångströms, and to low temperatures.

Under certain conditions, some systems might even show composite kinetics, for example niobium oxidising in air at about 1000 °C initially conforms to the parabolic law but later becomes linear, i.e. the rate becomes constant at long times.

Discontinuous methods of assessment of reaction kinetics

In this case, the specimen is weighed and measured and it is then exposed to the conditions of high temperature oxidation for a given time, removed, and reweighed. Using a bromine solution, which attacks the metal at the scale–metal interface, the oxide scale can be removed for detailed examination. Assessment of the extent of reaction may be carried out quite simply either by noting the mass gain of the oxidised specimen, which is the mass of oxygen taken into the scale, or by noting the mass loss of the stripped specimen, which is equivalent to the amount of metal taken up in scale formation, so long as the scale can be removed cleanly and no subscale or internal attack is involved. Alternatively, the changes in specimen dimensions may be measured. As mentioned before, these techniques yield only one point per specimen with the disadvantages that (a) many specimens are needed to plot fully the kinetics of the reaction, (b) due to experimental variations the results from each specimen may not be equivalent, and (c) the progress of the reaction between the points is not observed. However, they have the obvious advantage that the techniques, and the apparatus required, are extremely simple.

Continuous methods of assessment of reaction kinetics

These methods fall into two types; those which monitor mass gain and those which monitor gas consumption.

Mass gain methods

The simplest method of continuous monitoring is using a spring balance. In this case, the specimen is suspended from a sensitive spring whose extension is measured, using a cathetometer, as the specimen gains mass due to oxidation, thus giving a semi-continuous monitoring of the reaction. Apparatus suitable for this is shown diagrammatically in Fig. 1.1 which is self-explanatory. An important feature is the design of the upper suspension point, shown here as a hollow glass tube which also acts as a gas outlet. The tube can be twisted, raised, or lowered to facilitate accurate placing of the specimen and alignment of the spring. A suspension piece is rigidly fastened to the glass tube and provides a serrated horizontal support for the spring whose suspension point may thus be adjusted in the horizontal plane. These refinements are required since alignment between the pyrex tube containing the spring and the furnace tube is never perfect and it is prudent to provide some means of adjustment of the spring position in order to ensure accurate placing of the specimen. It would be possible, of course, to equip a spring balance with a moving transformer which would enable the mass gain to be measured electrically and automatically recorded. Although the simple spring balance should be regarded as a semi-continuous method of assessment, it has the advantage that a complete reaction curve can be obtained from a single specimen. The disadvantage of the spring balance is that its usage gives rise to the need for a compromise between accuracy and sensitivity; for accuracy, a large specimen is required whereas, for sensitivity, a relatively fragile spring should be used. It is obviously not possible to use a fragile spring to carry a large specimen and so the accuracy which is obtained by this method is a matter of compromise between these two factors.

By far the most popular, most convenient, and, unfortunately, most expensive method of assessing oxidation reactions is to use the continuous automatic recording balance. Obviously the operator must decide precisely what is required in terms of accuracy and sensitivity from the balance. For straightforward oxidation experiments it is generally adequate to choose a balance with a load carrying ability of up to about 25 grams and an ultimate sensitivity of about 100 μg. This is not a particularly sensitive balance and it is rather surprising that many investigators use far more sophisticated, and expensive, semi-micro balances for this sort of work. In fact many problems arise from the use of very sensitive balances, together with small specimens, to achieve high accuracy. This technique is subject to errors due to changes in Archimedian buoyancy when the gas composition is changed or the temperature is altered. An error is also introduced due to a change in dynamic buoyancy when the gas flow rate over the specimen is altered. For the most trouble-free operation it is advisable to use a large specimen with a large flat surface area in conjunction with a balance of moderate accuracy. Figure 1.2 shows a schematic diagram of a continuously reading thermobalance used in the authors' laboratory.

Using the automatic recording balance, it is possible to get a continuous record of the reaction kinetics and, in this way, many details come to light which are hidden by other methods; for example, spalling of small amounts of

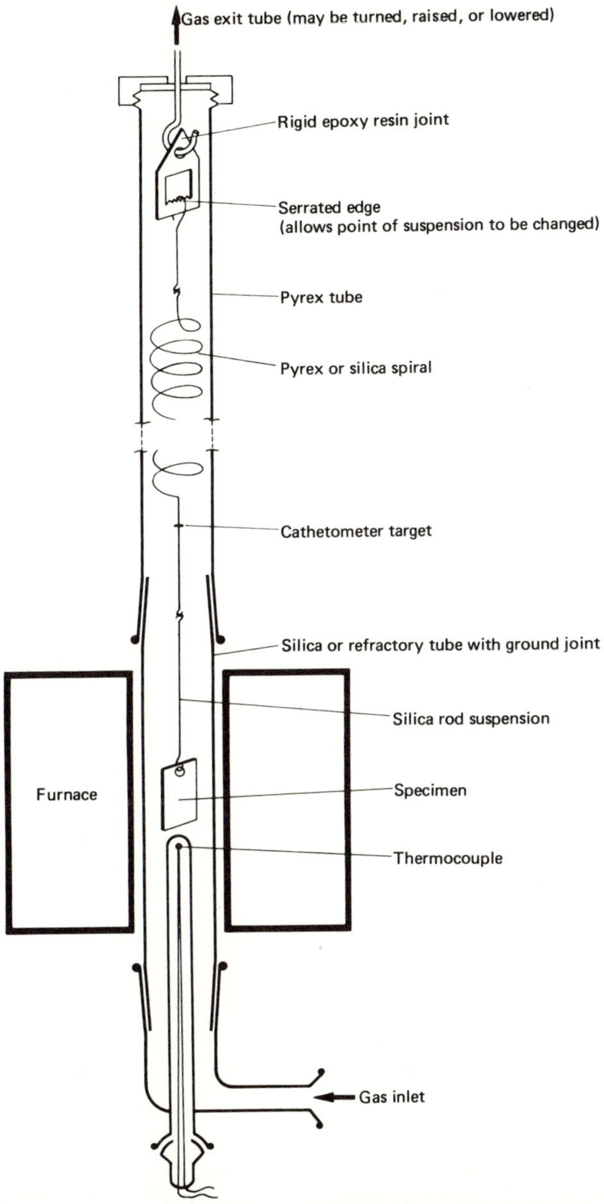

Fig. 1.1 Features of a simple spring balance (for advanced design see Mrowec, S. and Stoklosa, A., *J. Thermal. Anal.*, **2,** 73, 1970)

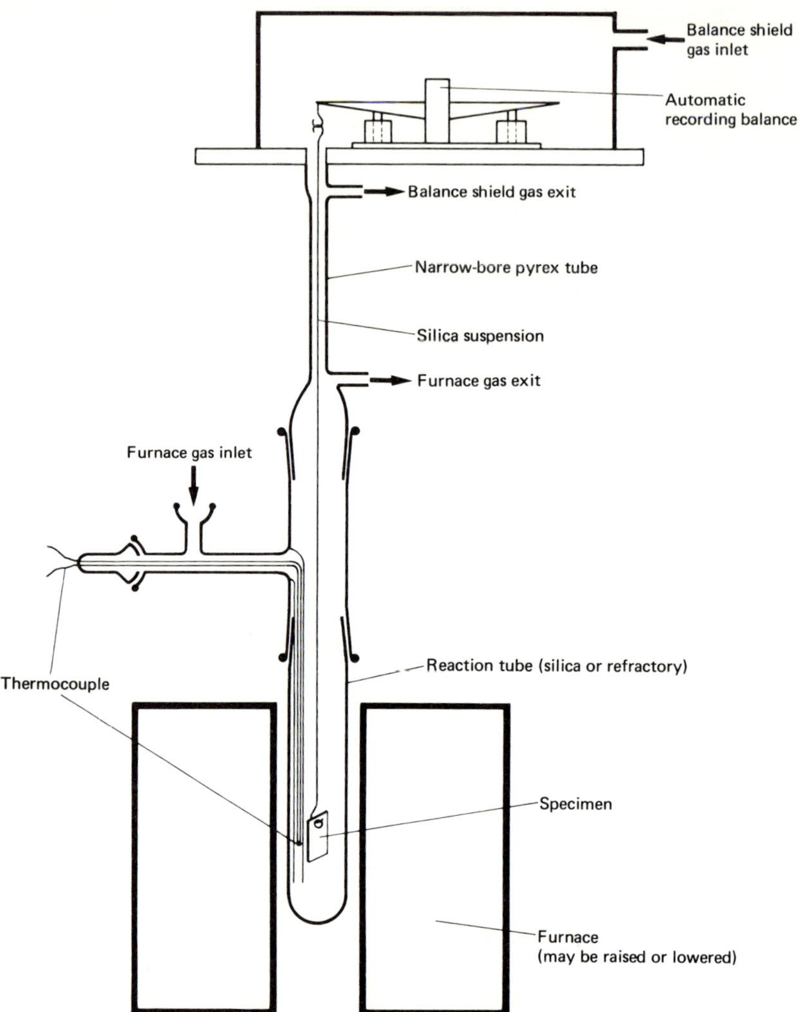

Fig. 1.2 Typical experimental arrangement for use with an automatic recording balance

the oxide layer is immediately detected as the balance registers the immediate weight loss, separation of the scale from the metal surface is recorded when the balance shows a slow reduction in rate well below rates which would be expected from the normal reaction laws, and cracking of the scale is indicated by immediate small increases in the rate of mass gain. These features are shown in Fig. 1.3. The correct interpretation of data yielded by the automatic recording balance requires skill and patience but, in almost every case, is well worthwhile.

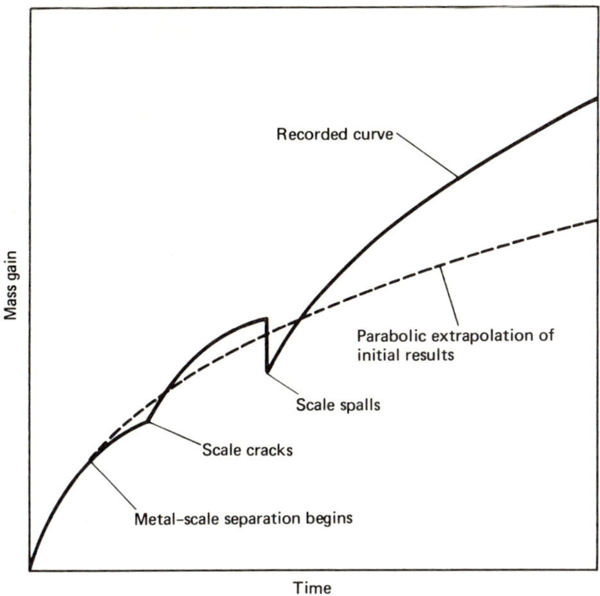

Fig. 1.3 Hypothetical mass gain versus time curve for an oxidation reaction showing, exaggerated, possible features that would be revealed by continuous monitoring, but missed by a discontinuous technique – leading to erroneous interpretation

Gas consumption methods

Continuous assessment based on measurement of the consumption of gas may be done in two ways. If the system is held at constant volume then the fall in pressure may be monitored, continuously or discontinuously. Unfortunately, allowing the pressure of the oxidant to vary appreciably during the reaction is almost certain to cause changes in the rate at which the reaction proceeds. The second method is to maintain the apparatus at constant pressure and to measure the fall in volume as oxidising gas is consumed; this can be done automatically and the results plotted using a chart recorder. This technique is only useful when the atmosphere is a pure gas, for example, pure oxygen. In the case of a mixture of gases such as air, the fall in volume would represent the amount of oxygen consumed but, since nitrogen is not being consumed, the partial pressure of oxygen would fall, leading to possible changes in oxidation rate. This problem can, to some extent, be overcome by using a very large volume of gas compared with the amount of gas used during the reaction; in this case there remains the difficulty of measuring a small change in a large volume. Other difficulties associated with the measurement of gas consumption arise from the sensitivity of these methods to variations in room temperature and atmospheric pressure.

Methods of examination

Initially the specimen should be examined by the naked eye, note being taken of whether the scale surface is flat, rippled, contains nodules, is cracked, or whether there is excessive attack on the edges or in the centres of the faces. This type of examination is important because subsequent examination under the microscope is usually carried out on a section at a specific place and it is important to know whether this is typical or if there are variations over the specimen surface. It is usually well worthwhile using the scanning electron microscope (SEM) before sectioning if the specimen can safely be mounted within the instrument and, as with all examination, if an area is seen which has significant features, photographs must be taken immediately. It is usually too late if one waits for a better opportunity. In addition, the modern SEM allows the chemical composition of surface features to be determined.

Having examined the scale, a little should be detached for X-ray diffraction work, for which care should be taken that the scale is detached from a representative position. If there are differences between the edges and the surface areas of the scale then samples should be taken from each part. X-ray diffraction is then carried out to give information on the phases present in the scales.

Metallographic examination involves polishing of a section and, under the polishing stresses, the scale is likely to fall away from the specimen. The specimen with its surrounding scale must, therefore, be mounted. By far the best type of mounting medium has been found to be the liquid epoxy resins. The process is very easy – a dish is greased, the specimen is positioned, and the resin is poured in. At this stage the dish should be evacuated in a desiccator, left under vacuum for a few minutes, and then atmospheric pressure reapplied. This has the effect of removing air from crevices caused by cracks in the scale and on reapplying pressure the resin is forced into cavities left by evacuation. This sort of mounting procedure requires several hours for hardening, preferably being left in a warm place overnight, but the good support which is given to the fragile scales is well worth the extra time and effort.

Polishing

It is important at this stage to realise that polishing is carried out for the oxides and not for the metal. A normal metallurgical polish will not reveal the full features of the oxide. Normally, a metal is polished using successive grades of abrasive paper – the time spent on each paper being equivalent to the time required to remove the scratches left by the previous paper. If this procedure is carried out it will almost certainly produce scales which are highly porous. It is important to realise that, due to their friability, scales are damaged to a greater depth than that represented by the scratches on the metal and, when going to a less abrasive paper, it is important to polish for correspondingly longer in order to undercut the depth of damage caused by the previous abrasive paper, i.e. polishing should be carried out for much longer times on each paper than would be expected from examination of the metal surface. If this procedure is carried out, scales which are nicely compact and free from porosity will be revealed. The polished specimens may be examined using conventional optical metallography or scanning electron microscopy.

Finally, it may be found necessary to etch a specimen to reveal detail in either the metal or the oxide scale. This is done following standard metallographic preparation procedure with the exception that thorough and lengthy rinsing in the neutral solvent (alcohol) must be carried out. This is necessary due to the existence of pores and cracks in the oxide, especially at the scale–metal interface, that retain the etchant by capillary action. Thus, extensive soaking in rinse baths is advisable, followed by rapid drying under a hot air stream, if subsequent oozing and staining are to be avoided.

It is important to emphasise that proper investigation of oxidation mechanisms involves as many of the above techniques of observing kinetics and morphologies as feasible, and careful combination of the results.

2
Thermodynamic fundamentals

Introduction

A sound understanding of high temperature corrosion reactions requires the determination of whether or not a given component in a metal or alloy can react with a given component from the gas phase or another condensed phase and to rationalise observed products of the reactions. In practice, the corrosion problems to be solved are often complex involving the reaction of multicomponent alloys with gases containing two or more reactive components. The situation is often complicated by the presence of liquid or solid deposits which form either by condensation from the vapour or impaction of particulate matter. An important tool in the analysis of such problems is, of course, equilibrium thermodynamics which, although not predictive, allows one to ascertain which reaction products are possible, whether or not significant evaporation or condensation of a given species is possible, the conditions under which a given reaction product can react with a condensed deposit, etc. The complexity of the corrosion phenomena usually dictates that the thermodynamic analysis be represented in graphical form.

The purpose of this chapter is to review the thermodynamic concepts pertinent to gas–metal reactions and then to describe the construction of the thermodynamic diagrams most often used in corrosion research and to present illustrative examples of their application. The types of diagrams discussed are

(a) Gibbs free energy versus composition diagrams and activity versus composition diagrams which are used for describing the thermodynamics of solutions.
(b) Standard free energy of formation versus temperature diagrams which allow the thermodynamic data for a given class of compounds, oxides, sulphides, carbides, etc., to be presented in a compact form.
(c) Vapour species diagrams which allow the vapour pressures of compounds to be presented as a function of a convenient variable such as pressure of a gaseous component.
(d) Two-dimensional, isothermal stability diagrams which map the stable phases in systems involving one metallic and two reactive, non-metallic components.

(e) Two-dimensional, isothermal stability diagrams which map the stable phases in systems involving two metallic components and one reactive, non-metallic component.
(f) Three-dimensional, isothermal stability diagrams which map the stable phases in systems involving two metallic and two reactive, non-metallic components.

Basic thermodynamics

The question of whether or not a reaction can occur is answered by the second law of thermodynamics. Since the conditions most often encountered in high temperature reactions are constant temperature and pressure, the second law is most conveniently written in terms of the Gibbs free energy (G').

$$G' = H' - TS' \tag{2.1}$$

where H' is the enthalpy and S' the entropy of the system. Under these conditions the second law states

$\Delta G' < 0$ spontaneous reaction expected
$\Delta G' = 0$ equilibrium
$\Delta G' > 0$ thermodynamically impossible process

For a chemical reaction, e.g.

$$a\mathrm{A} + b\mathrm{B} = c\mathrm{C} + d\mathrm{D} \tag{2.2}$$

$\Delta G'$ is expressed as

$$\Delta G' = \Delta G^{\ominus} + RT\ln\left(\frac{a_{\mathrm{C}}^{c}\ a_{\mathrm{D}}^{d}}{a_{\mathrm{A}}^{a}\ \ a_{\mathrm{B}}^{b}}\right) \tag{2.3}$$

where $\Delta G^{\ominus}$ is the free energy change when all species are present in their standard states and a is the thermodynamic activity which describes the deviation from the standard state for a given species and may be expressed for a given species i as

$$a_{\mathrm{i}} = \frac{p_{\mathrm{i}}}{p_{\mathrm{i}}^{\ominus}} \tag{2.4}$$

where p_{i} is either the vapour pressure over a condensed species or the partial pressure of a gaseous species and $p_{\mathrm{i}}^{\ominus}$ is the same quantity corresponding to the standard state of i. Expressing a_{i} by equation 2.4 requires the reasonable approximation of ideal gas behaviour at the high temperatures and relatively low pressures usually encountered. The standard free energy change is expressed for a reaction such as equation 2.2 by

$$\Delta G^{\ominus} = c\Delta G_{\mathrm{C}}^{\ominus} + d\Delta G_{\mathrm{D}}^{\ominus} - a\Delta G_{\mathrm{A}}^{\ominus} - b\Delta G_{\mathrm{B}}^{\ominus} \tag{2.5}$$

where $\Delta G_{\mathrm{C}}^{\ominus}$, etc., are standard free energies of formation which may be obtained from tabulated values. (A selected list of references for thermodynamic data such as $\Delta G^\ominus$ is included at the end of this chapter.) For the special case of equilibrium ($\Delta G' = 0$) equation 2.3 reduces to

$$\Delta G^\ominus = -RT \ln \left(\frac{a_C^c \, a_D^d}{a_A^a \, a_B^b} \right)_{eq} \tag{2.6}$$

The bracketed term is called the equilibrium constant (K) and is used to describe the equilibrium state of the reacting system.

Construction and use of thermodynamic diagrams

(a) Gibbs free energy versus composition diagrams and activity versus composition diagrams

The equilibrium state of a system at constant temperature and pressure is characterised by a minimum in the Gibbs free energy of the system. For a multicomponent, multiphase system, the minimum free energy corresponds to uniformity of the chemical potential (μ) of each component throughout the system. For a binary system, the molar free energy (G) and chemical potentials are related by

$$G = (1 - X)\mu_A + X\mu_B \tag{2.7}$$

where μ_A and μ_B are the chemical potentials of components A and B respectively and $(1 - X)$ and X are the mole fractions of A and B. Figure 2.1(a) is a plot of G versus composition for a single phase which exhibits

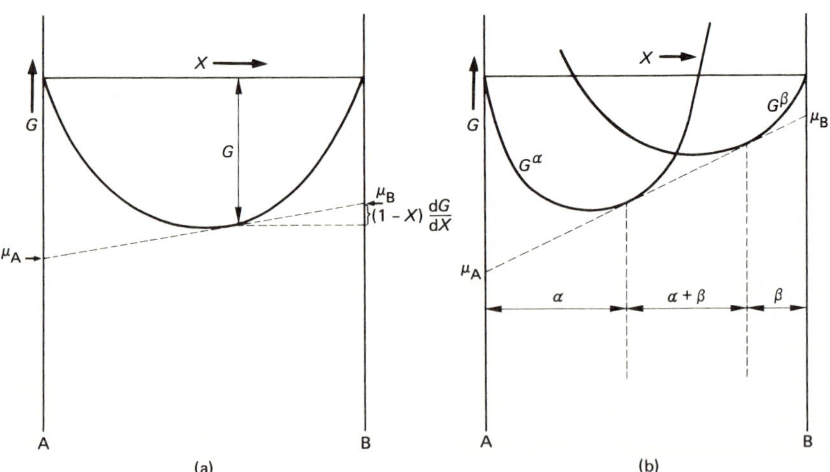

Fig. 2.1 Free energy versus composition diagrams for (a) one phase, and (b) two coexisting phases

simple solution behaviour. It can readily be shown[1,2], starting with equation 2.7, that for a given composition

$$\mu_A = G - X\frac{dG}{dX} \tag{2.8}$$

$$\mu_B = G + (1 - X)\frac{dG}{dX} \tag{2.9}$$

which, as indicated in Fig. 2.1(a), mean that a tangent drawn to the free energy curve has intercepts on the ordinate at $X = 0$ and $X = 1$ which correspond to the chemical potentials of A and B respectively. Alternatively, if the free energy of the unmixed components in their standard states, $(1 - X)\mu_A^\ominus + X\mu_B^\ominus$, is subtracted from each side of equation 2.7, the result is the molar free energy of mixing

$$\Delta G^M = (1 - X)(\mu_A - \mu_A^\ominus) + X(\mu_B - \mu_B^\ominus) \tag{2.10}$$

and a plot of ΔG^M versus X may be constructed in which case the tangent intercepts are $(\mu_A - \mu_A^\ominus)$ and $(\mu_B - \mu_B^\ominus)$ respectively. These quantities, the partial molar free energies of mixing, are directly related to the activities of the given component through

$$\mu_A - \mu_A^\ominus = RT \ln a_A \tag{2.11}$$

$$\mu_B - \mu_B^\ominus = RT \ln a_B \tag{2.12}$$

Figure 2.1(b) shows the free energy versus composition diagram for two phases. Application of the lever rule shows that the free energy for a mixture of the two phases is given by the point on the chord drawn between the two individual free energy curves where it intersects the bulk composition. Clearly, for any bulk composition in the central portion of Fig. 2.1(b), a two phase mixture will have a lower free energy than a single solution corresponding to either phase and the lowest free energy of the system, i.e. the lowest chord, obtains when the chord is in fact a tangent to both free energy curves. This 'common tangent construction' indicates X_1 and X_2 to be the equilibrium compositions of the two coexisting phases and the intercepts of the tangent at $X = 0$ and $X = 1$ show the construction is equivalent to the equality of chemical potentials, i.e.

$$\mu_A^\alpha = \mu_A^\beta \tag{2.13}$$

$$\mu_B^\alpha = \mu_B^\beta \tag{2.14}$$

Equations 2.11 and 2.12 indicate that similar information may be expressed in terms of activity. Figure 2.2 shows activity versus composition plots for a system with a free energy curve such as that in Fig. 2.1(a); with Fig. 2.2(a)

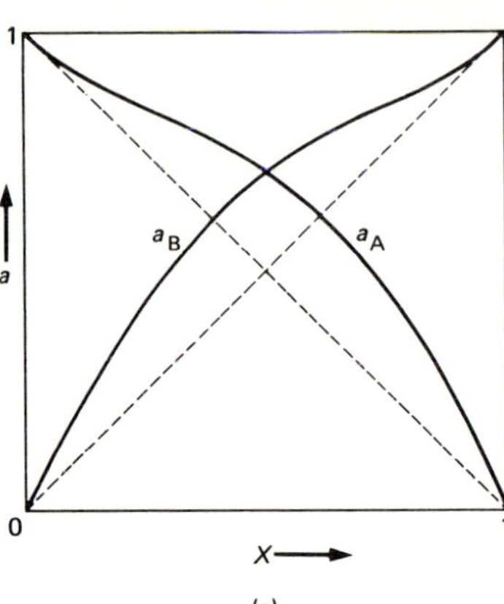

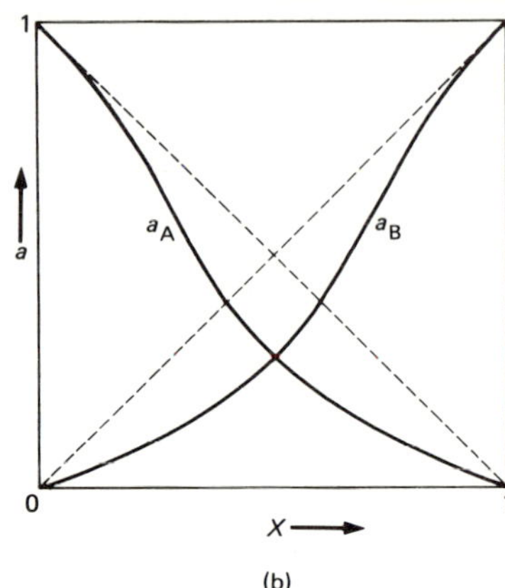

Fig. 2.2 Activity versus composition diagrams for a solution exhibiting (a) positive deviation from Raoult's law and (b) negative deviation from Raoult's law

corresponding to a slightly positive deviation from Raoult's law ($a > X$) and Fig. 2.2(b) corresponding to a slightly negative deviation ($a < X$). For cases where a two phase mixture is stable, Fig. 2.1(b), equations 2.13 and 2.14 indicate the activities are constant across the two phase field. Figure 2.3 illustrates the above points and their relationship to the conventional phase diagram for a hypothetical binary eutectic system. For a comprehensive treatment of free energy diagrams and their application to phase transformations the reader is referred to the excellent article by Hillert[3].

Clearly, if the variation of G or ΔG^M with composition for a system can be established, either by experimental activity measurements or by calculations using solution models, the phase equilibria can be predicted for a given temperature and composition. Considerable effort has been spent in recent years on computer calculations of phase diagrams, particularly for refractory metal systems (see for example reference 4). Furthermore, the concepts need not be limited to binary elemental systems. For example, ternary free energy behaviour may be described on a three-dimensional plot the base of which is a Gibbs triangle and the vertical axis represents G (or ΔG^M). In this case, the equilibrium state of the system is described using a 'common tangent plane', rather than a common tangent line, but the principles are unchanged. Systems with more than three components may be treated but graphical interpretation becomes more difficult. Also, the above concepts may be applied to pseudo-binary systems where the components are taken as compounds of fixed stoichiometry. Figure 2.4 shows the activity versus composition plot[5] for such a system, Fe_2O_3–Mn_2O_3. This type of diagram may be constructed for any of the systems encountered in high temperature corrosion, e.g. sulphides, oxides, sulphates, etc., and is useful for interpreting corrosion products and deposits formed on metallic or ceramic components.

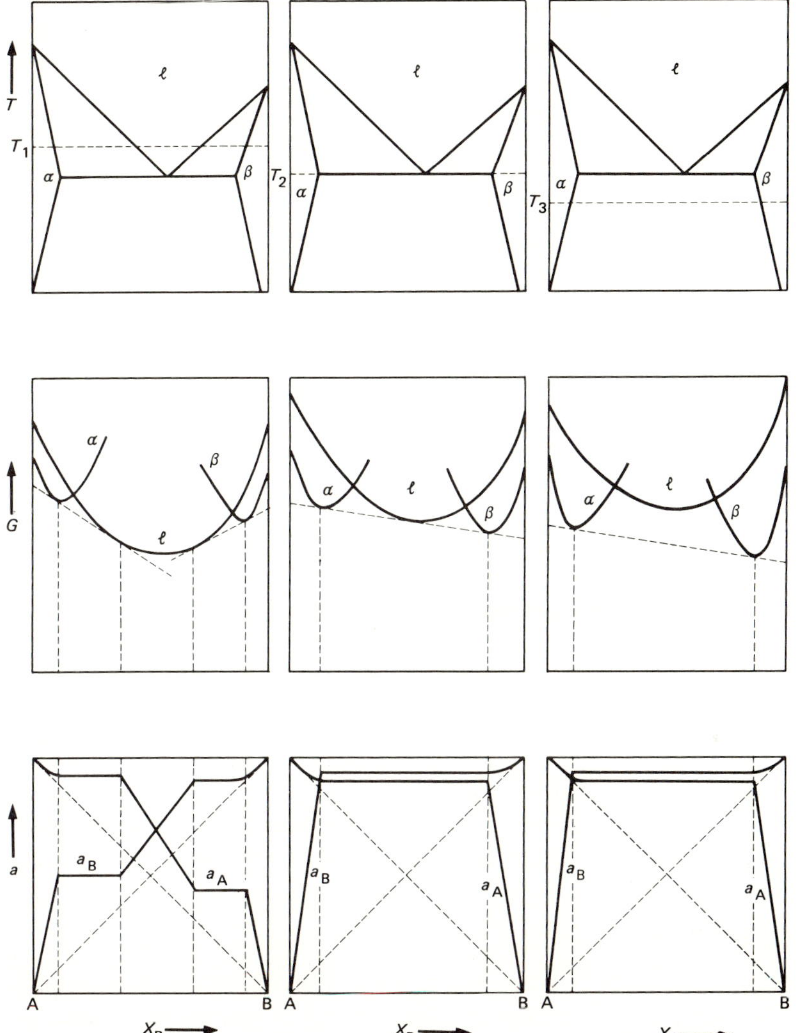

Fig. 2.3 Free energy versus composition and activity diagrams for a hypothetical eutectic system

(b) Standard free energy of formation versus temperature diagrams

Often determination of the conditions under which a given corrosion product is likely to form is required, e.g. in selective oxidation of alloys. In this regard, Ellingham diagrams, i.e. plots of the standard free energy of formation ($\Delta G^{\ominus}$) versus temperature for the compounds of a type, e.g. oxides, sulphides, carbides, etc., are useful in that they allow comparison of

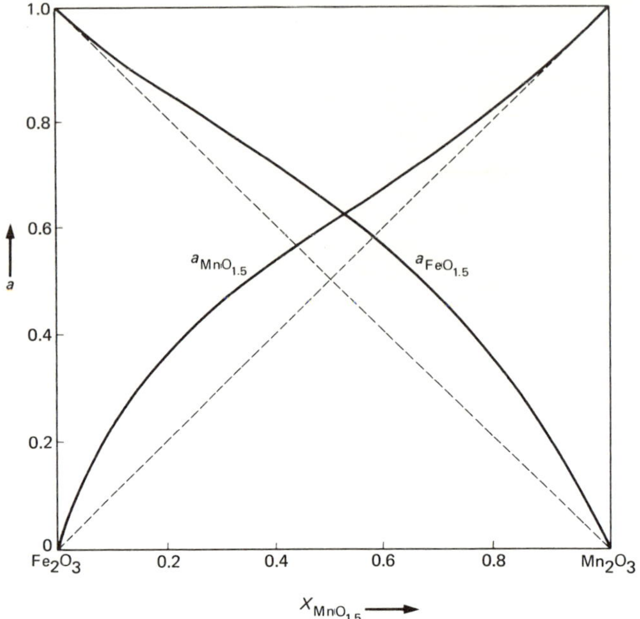

Fig. 2.4 Activity versus composition diagram for the Fe_2O_3–Mn_2O_3 system at 1300 °C (see Sticher and Schmalzried[5])

the relative stabilities of each compound. Figure 2.5 is such a plot for many simple oxides. The values of $\Delta G^\ominus$ are expressed as kilojoules per mole O_2 so the stabilities of various oxides may be compared directly, i.e. the lower the position of the line on the diagram the more stable is the oxide. Put another way, for a given reaction

$$M + O_2 = MO_2 \tag{2.15}$$

if the activities of M and MO_2 are taken as unity. Equation 2.6 may be used to express the oxygen partial pressure at which the metal and oxide coexist, i.e. the dissociation pressure of the oxide.

$$p_{O_2}^{M/MO_2} = \exp\frac{\Delta G^\ominus}{RT} \tag{2.16}$$

Naturally, in considering alloy oxidation, the activity of the metal and oxide must also be taken into account, i.e.

$$p_{O_2}^{eq} = \frac{a_{MO_2}}{a_M}\exp\frac{\Delta G^\ominus}{RT} = \frac{a_{MO_2}}{a_M}p_{O_2}^{M/MO_2} \tag{2.17}$$

The values of $p_{O_2}^{M/MO_2}$ may be obtained directly from the oxygen nomograph on the diagram by drawing a straight line from the origin marked 0 through

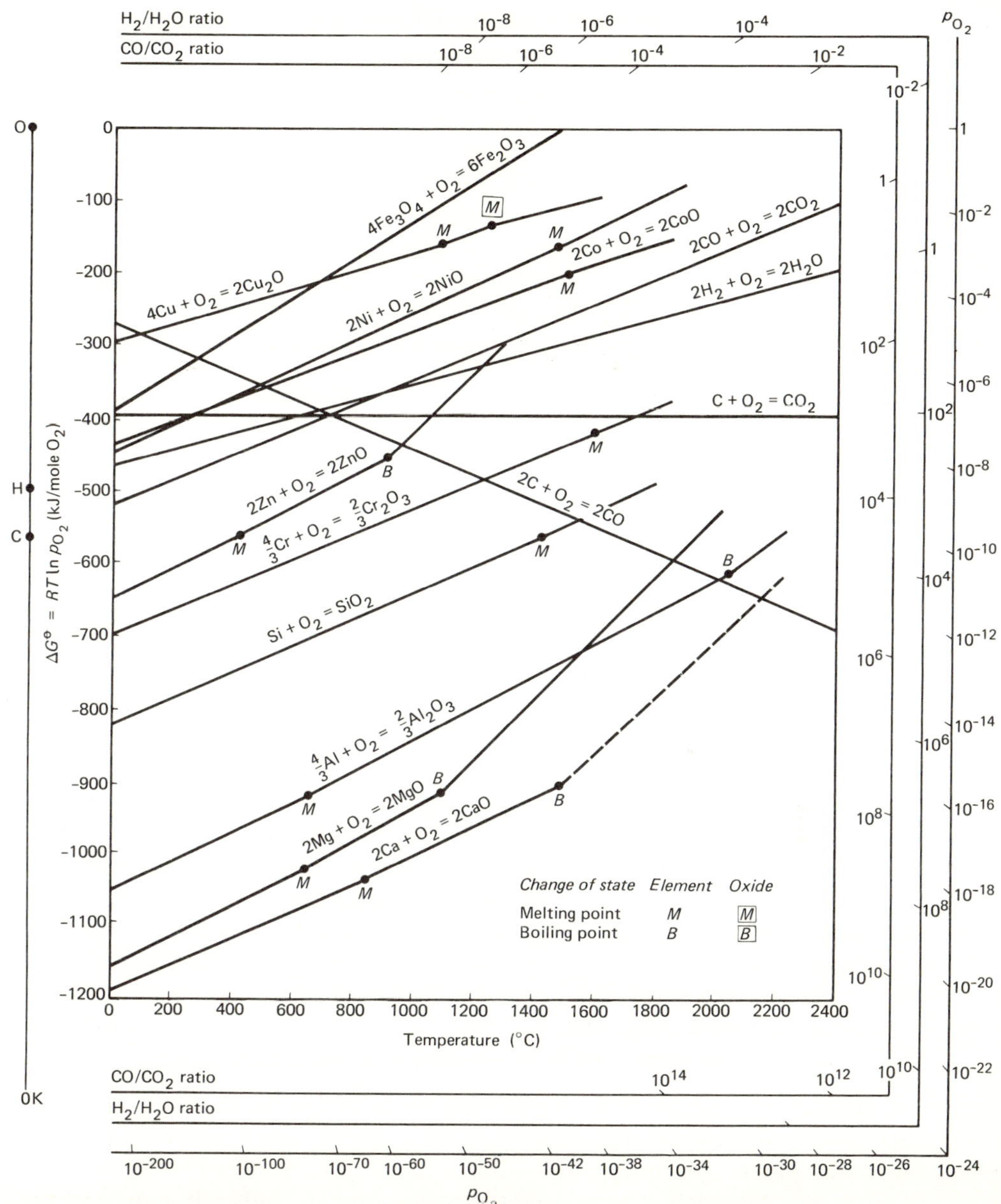

Fig. 2.5 Standard free energy of formation of selected oxides as a function of temperature

the free energy line at the temperature of interest and reading the oxygen pressure from its intersection with the scale at the right side labelled p_{O_2}. Values for the pressure ratio H_2/H_2O for equilibrium between a given metal and oxide may be obtained by drawing a similar line from the point marked H to the scale labelled H_2/H_2O ratio and values for the equilibrium CO/CO_2 ratio may be obtained by drawing a line from point C to the scale CO/CO_2 ratio. The reader is referred to Gaskell[2], Chapter 10, for a more detailed discussion of the construction and use of Ellingham diagrams for oxides.

Ellingham diagrams may, of course, be constructed for any class of compounds. Shatynski has published Ellingham diagrams for sulphides[6] and carbides[7]. Ellingham diagrams for nitrides and chlorides are presented in Chapter 14 of reference 1.

(c) Vapour species diagrams

The vapour species which form in a given high temperature corrosion situation often have a strong influence on the rate of attack, the rate generally being accelerated when volatile corrosion products form. Gulbransen and Jansson[8] have shown that metal and volatile oxide species are important in the kinetics of high temperature oxidation of carbon, silicon, molybdenum, and chromium. Six types of oxidation phenomena were identified

(1) at low temperature, diffusion of oxygen and metal species through a compact oxide film,
(2) at moderate and high temperatures, a combination of oxide film formation and oxide volatility,
(3) at moderate and high temperatures, the formation of volatile metal and oxide species at the metal–oxide interface and transport through the oxide lattice and mechanically formed cracks in the oxide layer,
(4) at moderate and high temperatures, the direct formation of volatile oxide gases,
(5) at high temperature, the gaseous diffusion of oxygen through a barrier layer of volatilised oxides,
(6) at high temperature, spalling of metal and oxide particles.

Sulphidation reactions follow a similar series of kinetic phenomena as has been observed for oxidation. Unfortunately, few studies have been made of the basic kinetic phenomena involved in sulphidation reactions at high temperature. Similarly, the volatile species in sulphate and carbonate systems are important in terms of evaporation/condensation phenomena involving these compounds on alloy or ceramic surfaces. Perhaps the best example of this behaviour is the rapid degradation of protective scales on many alloys, termed 'hot corrosion', which occurs when Na_2SO_4 or other salt condenses on the alloy.

The diagrams most suited for presentation of vapour pressure data in oxide systems are $\log p_{M_xO_y}$, for a fixed T, versus $\log p_{O_2}$ and $\log p_{M_xO_y}$, for a fixed p_{O_2}, versus $1/T$ diagrams. The principles of the diagrams will be illustrated for the Cr–O system at 1250 K. Only one condensed oxide, Cr_2O_3, is formed under conditions of high temperature oxidation. Thermochemical data are available for $Cr_2O_3(s)$[9], Cr(g)[10], and the three gaseous oxide species CrO(g), $CrO_2(g)$, and $CrO_3(g)$[11]. These data may be expressed as standard free energies of formation ($\Delta G^{\ominus}$) or, more conveniently, as $\log K_p$ defined by

$$\log K_p = \frac{-\Delta G^{\ominus}}{2.303RT} \tag{2.18}$$

The appropriate data at 1250 K are

Species		*Log K_p*
Cr_2O_3	(s)	33.95
Cr	(g)	−8.96
CrO	(g)	−2.26
CrO_2	(g)	4.96
CrO_3	(g)	8.64

The vapour pressures of species in equilibrium with Cr metal must be determined for low oxygen pressures and those of species in equilibrium with Cr_2O_3 at high oxygen pressures. The boundary between these regions is the oxygen pressure for Cr/Cr_2O_3 equilibrium obtained from the equilibrium

$$2Cr(s) + \tfrac{3}{2}O_2(g) = Cr_2O_3(s) \tag{2.19}$$

for which

$$\log p_{O_2} = -\tfrac{2}{3} \log K_p^{Cr_2O_3} = -22.6 \tag{2.20}$$

This boundary is represented by the vertical line in Fig. 2.6. At lower oxygen pressures the pressure of Cr(g) is independent of p_{O_2}

$$Cr(s) = Cr(g) \tag{2.21}$$

and

$$\log p_{Cr} = \log K_p^{Cr(g)} = -8.96 \tag{2.22}$$

For oxygen pressures greater than Cr/Cr_2O_3 equilibrium the vapour pressure of Cr may be obtained from the equilibrium

$$Cr_2O_3(s) = 2Cr(g) + \tfrac{3}{2}O_2(g) \tag{2.23}$$

with the equilibrium constant determined from

$$\log K_{23} = 2 \log K_p^{Cr(g)} - \log K_p^{Cr_2O_3} = 51.9 \tag{2.24}$$

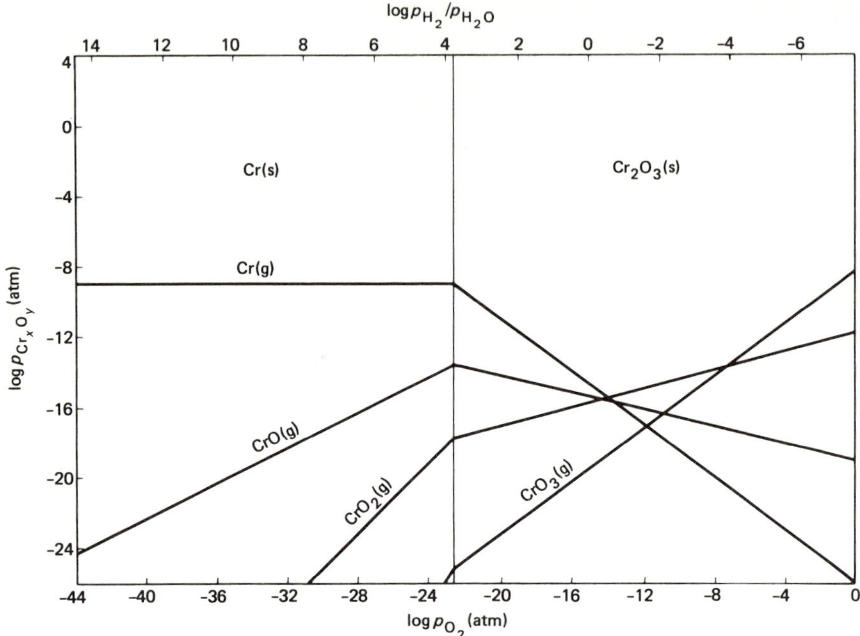

Fig. 2.6 Cr–O system volatile species at 1250 K

Therefore

$$2 \log p_{Cr} + \tfrac{3}{2} \log p_{O_2} = 51.9 \qquad (2.25)$$

or

$$\log p_{Cr} = 25.95 - \tfrac{3}{4} \log p_{O_2} \qquad (2.26)$$

The vapour pressure of Cr is then expressed in Fig. 2.6 by plotting equations 2.22 and 2.26. The lines corresponding to the other vapour species are obtained in the same manner. For example, for CrO_3 at oxygen pressures below Cr/Cr_2O_3 equilibrium

$$Cr(s) + \tfrac{3}{2}O_2(g) = CrO_3(g) \qquad (2.27)$$

and

$$\log K_{27} = \log K_p^{CrO_3(g)} = 8.64 \qquad (2.28)$$

from which we find

$$\log p_{CrO_3} = 8.64 + \tfrac{3}{2} \log p_{O_2} \qquad (2.29)$$

The other lines in Fig. 2.6 are obtained by writing similar equilibrium equations. Figure 2.6 shows that significant vapour pressures of Cr are developed at low p_{O_2}, e.g. at the alloy–scale interface of a Cr_2O_3-forming alloy, and that very large pressures of CrO_3 are developed at high p_{O_2}. The latter are responsible for a phenomenon whereby Cr_2O_3 scales are thinned by vapour losses during oxidation at high p_{O_2}, particularly for high gas flow rates. Figure 2.7 indicates the temperature dependence of the various equilibria. Similar diagrams may be constructed for other oxides, sulphides, halides, etc.

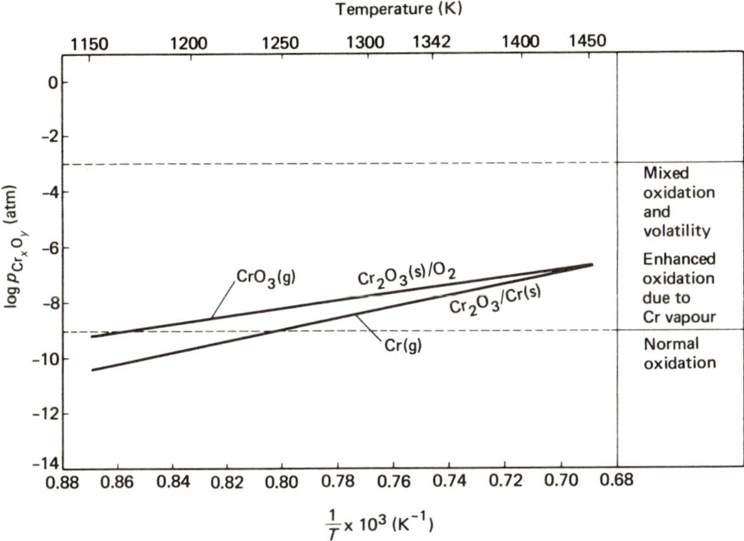

Fig. 2.7 Cr–O system volatile species versus temperature

(d) Two-dimensional, isothermal stability diagrams one metallic and two non-metallic components

When a metal reacts with a gas containing more than one oxidant, a number of different phases may form depending on both thermodynamic and kinetic considerations. Isothermal stability diagrams, usually constructed with the logarithmic values of the activities or partial pressures of the two non-metallic components as the coordinate axes, are useful in interpreting the condensed phases which form. To illustrate the construction of this type of diagram, the Ni–S–O system at 1250 K will be considered. The appropriate data[11] are

Species		*Log* K_p
NiO	(s)	5.34
NiS_y	(l)	$\log p_{S_2} = -6.87$
$NiSO_4$	(s)	15.97
SO_2	(g)	11.314
SO_3	(g)	10.563
S	(l)	−0.869

The coordinates chosen to express the equilibria are log p_{S_2} versus log p_{O_2}. (Only one sulphide is considered for simplicity.)

The assumption is made that all the condensed species are at unit activity. This assumption places important limitations on the use of the diagrams for alloy systems as will be discussed in later sections. Consider first the Ni/NiO equilibrium

$$Ni(s) + \tfrac{1}{2}O_2(g) = NiO(s) \qquad (2.30)$$

and

$$\log p_{O_2} = -2 \log K_p^{NiO} = -10.68 \qquad (2.31)$$

resulting in the vertical phase boundary in Fig. 2.8. Similarly, for $NiS_y(l)$ $(y \approx 1)$

$$Ni(s) + \frac{y}{2} S_2(g) = NiS_y(l) \qquad (2.32)$$

and

$$\log p_{S_2} = -6.87 \qquad (2.33)$$

resulting in the horizontal boundary in Fig. 2.8. Equations 2.31 and 2.33, therefore, define the gaseous conditions over which metallic Ni is stable. In a similar manner, the other phase boundaries are determined. For example, for the $NiO/NiSO_4$ equilibrium

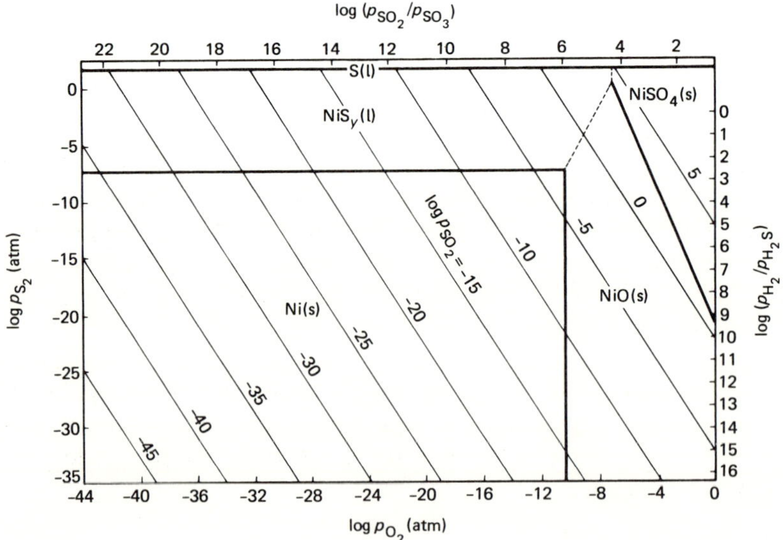

Fig. 2.8 Condensed phase equilibria for the Ni–O–S system at 1250 K

$$NiO(s) + \tfrac{1}{2}S_2(g) + \tfrac{3}{2}O_2(g) = NiSO_4 \tag{2.34}$$

and

$$\log K_{34} = \log K_p^{NiSO_4} - \log K_p^{NiO} = 10.63 \tag{2.35}$$

resulting in the equation for the phase boundary

$$\log p_{S_2} = -21.26 - 3 \log p_{O_2} \tag{2.36}$$

An additional feature of the diagrams are the superposed SO_2 isobars which define the S_2 and O_2 partial pressures which may obtain for a fixed SO_2 pressure. These isobars are obtained from the equilibrium

$$\tfrac{1}{2}S_2(g) + O_2(g) = SO_2(g) \tag{2.37}$$

with

$$\log K_{37} = \log K_p^{SO_2} = 11.314 \tag{2.38}$$

resulting in the equation

$$\log p_{S_2} = -22.628 + 2 \log p_{SO_2} - 2 \log p_{O_2} \tag{2.39}$$

The isobars appear as diagonal lines of slope −2. The horizontal line near the top of the diagram indicates the sulphur pressure at which S(l) will condense from the gas. This type of diagram will be used in later chapters to interpret the reactions involving metals in complex atmospheres.

When significant intersolubility exists between neighbouring phases on stability diagrams, solution regions may be included as shown in Fig. 2.9 taken from the work of Giggins and Pettit[12]. (Note the ordinate and abscissa are reversed from Fig. 2.8 and the temperature is slightly lower.) The phase boundaries in Fig. 2.9 were calculated from the equilibria, such as equations 2.31, 2.33, and 2.36, but the boundaries of the solution regions are calculated taking the activities of the essentially pure species as unity and of the minor species as some limiting value. In constructing Fig. 2.9, the activity of NiS was taken as 10^{-2} in equilibrium with essentially pure NiO and the activity of NiO was taken as 10^{-2} in equilibrium with essentially pure NiS. Similar approximations have been made for the other equilibria. Obviously experimental data regarding the extent of intersolubility of phases and activities in solution are essential to quantify such diagrams.

For some applications, different coordinate axes become more convenient for the construction of stability diagrams. For example, the phase equilibria for the Ni–S–O system in Fig. 2.8 may be equally well represented on a plot of $\log p_{O_2}$ versus $\log p_{SO_3}$ since specification of these two variables fixes $\log p_{S_2}$ because of the equilibrium

$$\tfrac{1}{2}S_2(g) + \tfrac{3}{2}O_2(g) = SO_3(g) \tag{2.40}$$

The equilibria of Fig. 2.8 are replotted in Fig. 2.10 using these coordinates.

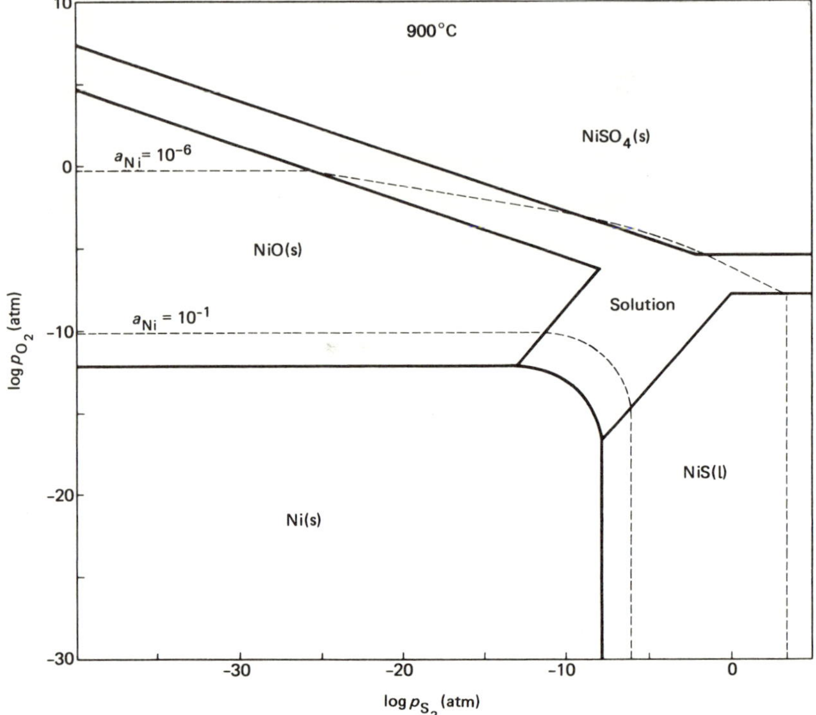

Fig. 2.9 Stability diagram for the Ni–O–S system where regions of solution were assumed to exist between NiO, NiS and $NiSO_4$ (from Giggins and Pettit[12])

This choice of coordinates is particularly useful in analysing hot corrosion situations such as those which occur when a metal or alloy is oxidised while covered with a deposit of sodium sulphate, since the components of Na_2SO_4 may be taken as Na_2O and SO_3. This phenomenon will be described in Chapter 7.

Stability diagrams of this type are not restricted to use with metal–sulphur–oxygen systems. As an example, Fig. 2.11 is a stability diagram for the Cr–C–O system at 1250 K, where the carbon activity is an important variable. The coordinates are log a_C versus log p_{O_2} and the construction and use of the diagram is exactly analogous to those for the metal–sulphur–oxygen diagrams. Applications of stability diagrams involving a number of systems have been discussed by Jansson and Gulbransen[13,14] and compilations of diagrams are available for metal–sulphur–oxygen systems[11] and for a number of metals in combinations of oxygen, sulphur, carbon, and nitrogen[15].

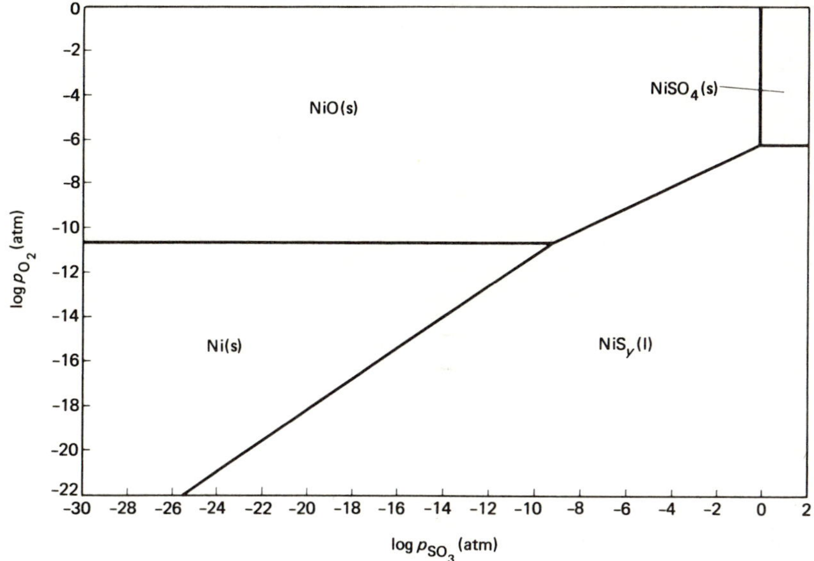

Fig. 2.10 Stability data for the Ni–O–S system at 1250 K expressed as $\log p_{O_2}$ versus $\log p_{SO_3}$

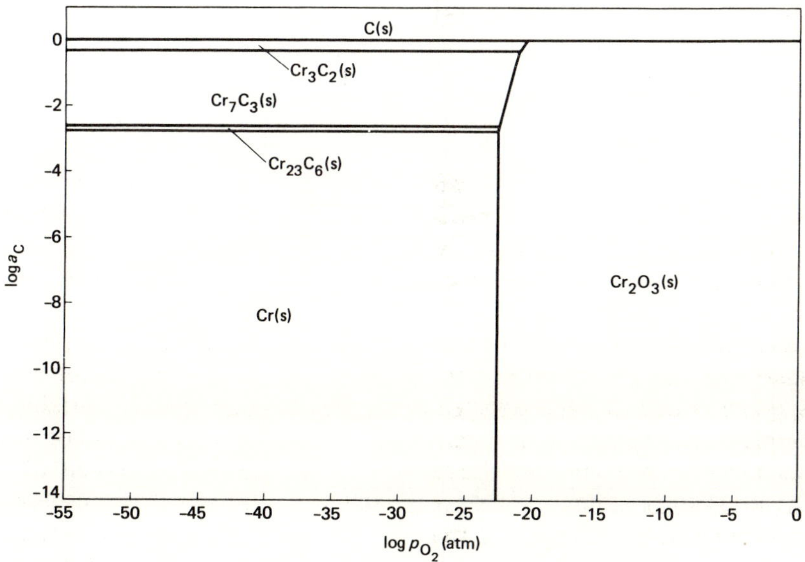

Fig. 2.11 Stability diagram for the Cr–C–O system at 1250 K

(e) Two-dimensional, isothermal stability diagrams two metallic components and one non-metallic component

In considering the reaction of even binary alloys with gases containing mixed oxidants, the thermodynamic description of the condensed phase equilibria becomes complicated by the addition of activity or composition of one metal component as a variable. However, before considering this problem, the condensed phase equilibria when only one oxidant is present must be considered. Figure 2.12, taken from the work of Sticher and Schmalzried[5], is the stability diagram for the Fe–Cr–O system at 1300 °C. The construction of this type of diagram requires the establishment of the free energy composition

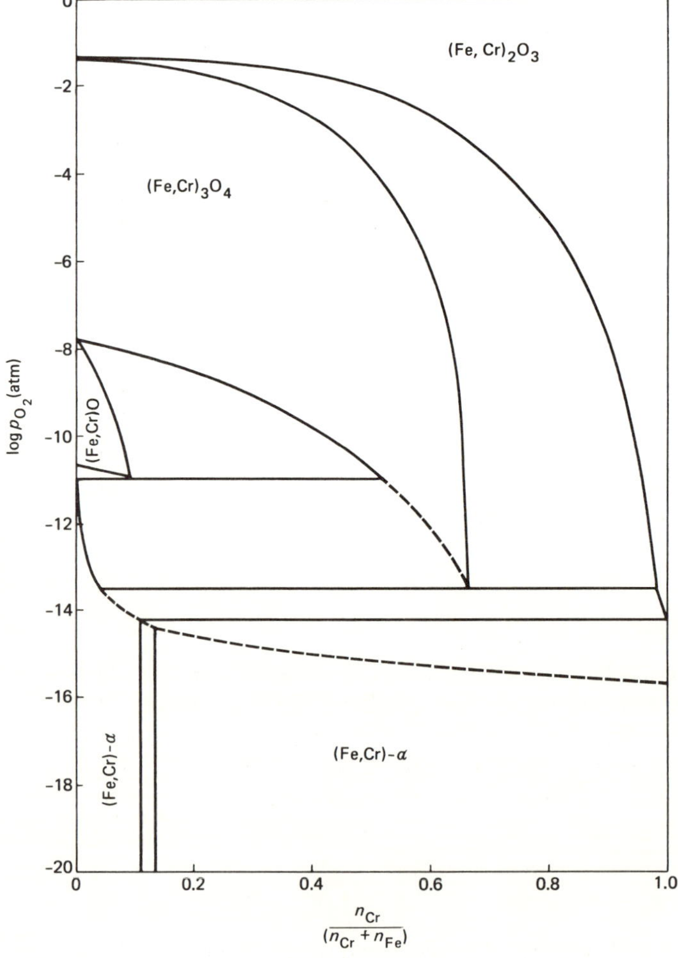

Fig. 2.12 Stability diagram for the Fe–Cr–O system at 1300 °C

behaviour (as discussed in Section (a)) for the system of interest and then the evaluation of the stability fields at a given temperature by minimising the free energy of the system, i.e. applying the common tangent construction. Few experimental data exist for the free energy composition relationships in mixed oxide or sulphide solutions, for example $(Fe, Cr)_3O_4$ or $(Fe, Cr)_{1-x} S_x$, so it is necessary to use various solution models to estimate these data[5,16]. Furthermore, few data exist for the free energies of formation of compounds of fixed stoichiometry, such as the sulphide spinel $FeCr_2S_4$, and these must be estimated[16].

The appearance of phases of variable composition and intermediate phases, such as spinels, as the mole fraction of a given metal is changed is significant since, as will be discussed in the next section, superposition of diagrams for the individual pure metals will neglect these phases. For example, considering the corrosion of Fe–Cr alloys in S–O atmospheres by superposing the Fe–S–O and Cr–S–O diagrams will neglect some of the important phase equilibria and should only be done with adequate justification.

(f) Three-dimensional, isothermal stability diagrams two metallic and two non-metallic components

The corrosion of alloys in gases containing mixed oxidants is a problem of major interest. The proper thermodynamic treatment of this type of problem requires a combination of the concepts presented in Sections (d) and (e). This is a difficult task and a lack of experimental data for ternary compounds has prevented significant calculations of this type. One exception is the work of Giggins and Pettit[12] in which a simplified three-dimensional diagram ($\log p_{S_2}$ versus $\log p_{O_2}$ versus $\log a_{Cr}$) has been constructed for the Ni–Cr–S–O system. This diagram is illustrated in Fig. 2.13.

Summary

The basic types of thermodynamic diagrams useful in interpreting the results of high temperature corrosion experiments have been discussed. While no attempt at a complete discussion of this subject has been made, it is hoped this chapter will facilitate a clear understanding of the high temperature corrosion phenomena to be described in the remainder of this book.

References

1 Darken, L. S. and Gurry, R. W., *Physical Chemistry of Metals*, McGraw-Hill, New York, 1953; Chapters 10 and 11

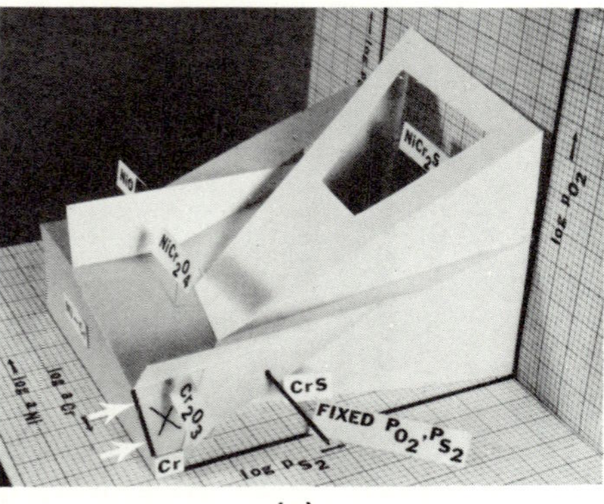

(a)

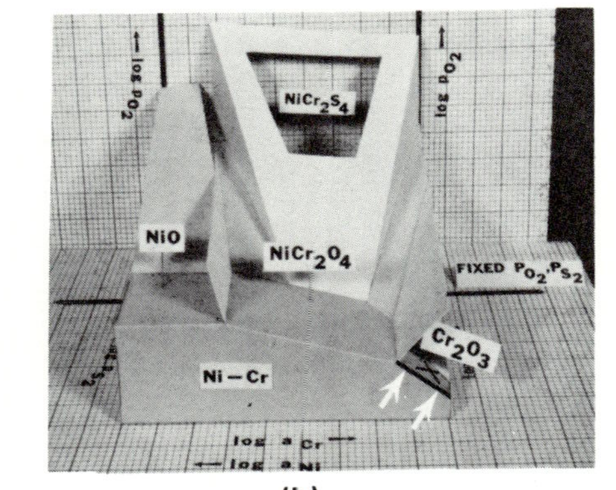

(b)

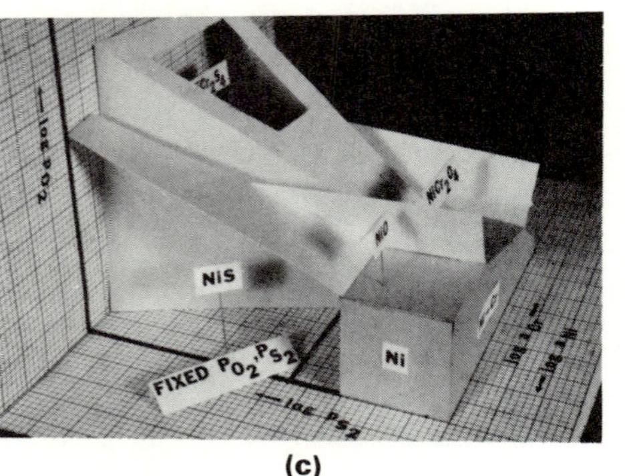

(c)

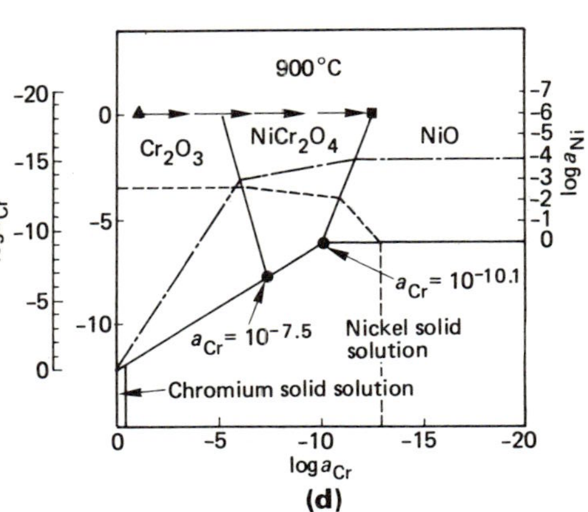

(d)

Fig. 2.13 (a), (b), and (c) three views of a schematic, three-dimensional diagram for the Ni–Cr–O–S system, the rod extending through the diagram shows the phases which are stable in a fixed gas composition. (d) stability diagram for the Ni–Cr–O system with isoactivity lines for chromium (– – –) and nickel (-.-); this diagram is an inverted version of the front plane of view (b) (from Giggins and Pettit[12])

2 Gaskell, D. R., *Introduction to Metallurgical Thermodynamics*, 2nd edition, McGraw-Hill, New York, 1981; Chapters 11 and 12

3 Hillert, M., The uses of Gibbs free energy–composition diagrams, in: *Lectures on the Theory of Phase Transformations*, ed. H. I. Aaronson, American Institute of Mining, Metallurgical, and Petroleum Engineers, 1975

4 Kaufman, L. and Bernstein, H., Computer calculations of refractory metal phase diagrams, in: *Phase Diagrams, Materials Science and Technology*, Volume 1, ed. A. M. Alper, Academic Press, New York, 1970

5 Sticher, J. and Schmalzried, H., *Zur geometrischen Darstellung thermodynamischer Zustandsgrossen in Mehrstoffsystemen auf Eisenbasis, Report*, Institute für Theoretische Hüttenkunde und Angewandte Physikalische Chemie der Technischen Universität Clausthal, 1975

6 Shatynski, S. R., *Oxid. Metals*, **11,** 307, 1977

7 Shatynski, S. R., *Oxid. Metals*, **13,** 105, 1979

8 Gulbransen, E. A. and Jansson, S. A., in: *Heterogeneous Kinetics at Elevated Temperatures*, ed. G. R. Belton and W. F. Worrell, Plenum Press, New York, 1970

9 Wicks, C. E. and Block, F. E., Thermodynamic properties of 65 elements, their oxides, halides, carbides, and nitrides, *Bureau of Mines, Bulletin 605,* U.S. Government Printing Office, Washington D.C., 1963

10 JANAF Thermochemical Tables, 1975 Supplement, *Reprint no. 60 from J. of Physical and Chemical Reference Data*, **4,** 1–175, 1975

11 Gulbransen, E. A. and Meier, G. H., Thermodynamic stability diagrams for condensed phases and volatility diagrams for volatile species over condensed phases in twenty metal–sulfur–oxygen systems between 1150 and 1450 K, *University of Pittsburgh, DOE Report on Contract no. DE-ACO1-79-ET-13547*, May 1979

12 Giggins, C. S. and Pettit, F. S., *Oxid. Metals*, **14,** 363, 1980

13 Gulbransen, E. A. and Jansson, S. A., General concepts of oxidation and sulfidation reactions: a thermochemical approach, in: *High Temperature Metallic Corrosion of Sulfur and Its Compounds*, ed. Z. A. Foroulis, Electrochem. Soc., New York, 1970

14 Jansson, S. A. and Gulbransen, E. A., Thermochemical considerations of high temperature gas–solid reactions, in: *High Temperature Gas–Metal Reactions in Mixed Environments*, ed. S. A. Jansson and Z. A. Foroulis, American Institute of Mining, Metallurgical, and Petroleum Engineers, New York, 1973

15 Hemmings, P. L. and Perkins, R. A., Thermodynamic phase stability diagrams for the analysis of corrosion reactions in coal gasification/combustion atmospheres, *FP-539, Research Project 716-1, Interim Report*, December 1977, prepared by Lockheed Palo Alto Research Laboratories for Electric Power Research Institute, Palo Alto, California

16 Jacob, K. T., Rao, D. Bhogeswara and Nelson, H. G., *Oxid. Metals*, **13**, 25, 1979

Selected thermodynamic data references

1 *JANAF Thermochemical Data*, including Supplements, Dow Chemical Co., Midland, Michigan; also *NSRDS-NBS 37*, U.S. Government Printing Office, Washington D.C., 1971; Supplements 1974, 1975 and 1978
2 Schick, H., *Thermodynamics of Certain Refractory Compounds*, 2 volumes, Academic Press, New York, 1966
3 Wicks, C. E. and Block, F. E., Thermodynamic properties of 65 elements, their oxides, halides, carbides, and nitrides, *Bureau of Mines, Bulletin 605*, U.S. Government Printing Office, Washington D.C., 1963
4 Hultgren, R., Orr, R. L., Anderson, P. D. and Kelley, K. K., *Selected Values of Thermodynamic Properties of Metals and Alloys*, Wiley, New York, 1967
5 Kubaschewski, O. and Alcock, C. B., *Metallurgical Thermochemistry*, 5th edition, Pergamon Press, Oxford, 1979
6 Wagman, D. D., and others, Selected values of chemical thermodynamic properties, elements 1 through 34, *NBS Technical Note 270-3*, U.S. Government Printing Office, Washington D.C., 1968; elements 35 through 53, *NBS Technical Note 270-4*, 1969
7 Stern, K. H. and Weise, E. L. *High Temperature Properties and Decomposition of Inorganic Salts. Part I. Sulfates*. National Standard Reference Data Series, National Bureau of Standards, no. 7, U.S. Government Printing Office, Washington D.C., 1966; *Part II. Carbonates*, 1969
8 Stull, D. R. and Sinke, G. C., *Thermodynamic Properties of the Elements*, Monograph 18, Advances in Chemistry, American Chemical Society, Washington D.C., 1956
9 Mills, K. C., *Thermodynamic Data for Inorganic Sulphides, Selenides, and Tellurides*, Butterworth, London, 1974
10 Dayhoff, M. O., Lippincott, E. R., Eck, R. V. and Nagarajan, G., *Thermodynamic Equilibrium in Prebiological Atmospheres of C, H, O, P, S, and Cl*, NASA SP-3040, 1964
11 Barin, I. and Knache, O., *Thermochemical Properties of Inorganic Substances*, Springer-Verlag, Berlin, Heidelberg, New York, 1973; also Supplement, 1977

3
Mechanisms of oxidation

Introduction

In this chapter the mechanisms by which ions and electrons transport through a growing oxide layer are described, using typical examples of both n-type and p-type semiconducting oxides. Using these concepts, the classical Wagner treatment of oxidation rates, controlled by ionic diffusion through the oxide layer, is presented showing how the parabolic rate constant is related to such fundamental properties as the partial ionic and electronic conductivities of the oxide and their dependence on the chemical potential of the metal or oxygen in the oxide. Finally, the mechanisms leading to observation of linear and logarithmic rate laws are discussed.

The discussion of defect structure has been simplified to that necessary to provide an understanding of oxidation mechanisms, however an extensive review of defect theory has been provided by Kröger[1] and the present state of knowledge concerning the defect structures of many oxides has been reviewed by Kofstad[2]. The reader is encouraged to consult these two sources for more detailed treatment of specific oxides.

Mechanisms of oxidation

From consideration of the equation

$$M(s) + \tfrac{1}{2}O_2(g) = MO(s)$$

it is obvious that the solid reaction product MO will separate the two reactants as shown below

M	MO	O_2
Metal	Oxide	Gas

In order for the reaction to proceed further, one or both reactants must penetrate the scale, i.e. either metal must be transported through the oxide to the oxide–gas interface and react there or oxygen must be transported to the oxide–metal interface and react there.

Immediately, therefore, the mechanisms by which the reactants may penetrate the oxide layer are seen to be an important part of the mechanism by which high temperature oxidation occurs. Precisely the same applies to the formation and growth of sulphides and other similar reaction products.

Transport mechanisms

Since all metal oxides and sulphides are ionic in nature it is not practicable to consider the transport of neutral metal or non-metal atoms through the reaction product. Several mechanisms are available to explain the transport of ions through ionic solids and these can be divided into mechanisms belonging to stoichiometric crystals and those belonging to non-stoichiometric crystals. In fact, although many compounds belong firmly to one or the other group, the correct view is probably that all of the possible defects are present to some extent or other in all compounds, but that in most cases certain types of defects predominate.

Strongly stoichiometric ionic compounds

Predominant defects in these compounds, which represent one limiting condition, are the Schottky and Frenkel defects.

Schottky defects Ionic mobility is explained by the existence of ionic vacancies. In order to maintain electroneutrality, it is necessary to postulate an equivalent number, or concentration, of vacancies on both cationic and anionic sub-lattices. This type of defect occurs in the alkali halides and is shown for KCl in Fig. 3.1. Since vacancies exist on both sub-lattices, it is to be expected that both anions and cations will be mobile.

K^+ Cl^- K^+ Cl^- K^+ Cl^-
Cl^- □ Cl^- K^+ Cl^- K^+
K^+ Cl^- K^+ □ K^+ Cl^-
Cl^- K^+ Cl^- K^+ Cl^- K^+
K^+ Cl^- K^+ □ K^+ Cl^-
Cl^- □ Cl^- K^+ Cl^- K^+
K^+ Cl^- K^+ Cl^- K^+ Cl^-

Fig. 3.1 Schottky defect in alkali halides

Frenkel defects The case where only the cation is mobile can be explained by assuming that the anion lattice is perfect but that the cation lattice contains cation vacancies and interstitials in equivalent concentrations to maintain electroneutrality for the whole crystal. This type of defect is found in the silver halides and is shown for AgBr in Fig. 3.2. The cation in this case is free to migrate over both vacancy and interstitial sites.

However, it is apparent that neither of these defects can be used to explain material transport during oxidation reactions because neither defect structure provides a mechanism by which electrons may migrate.

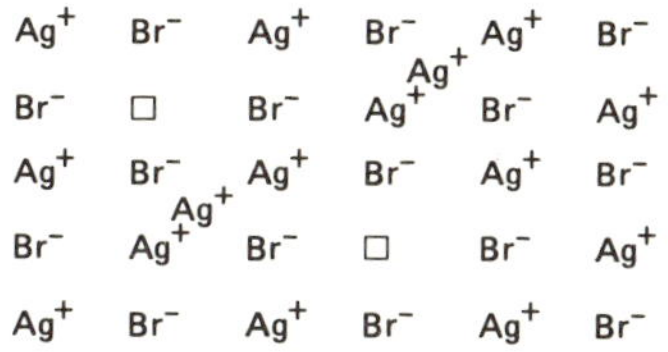

Ag+ Br− Ag+ Br− Ag+ Br−
Ag+
Br− □ Br− Ag+ Br− Ag+
Ag+ Br− Ag+ Br− Ag+ Br−
Ag+
Br− Ag+ Br− □ Br− Ag+
Ag+ Br− Ag+ Br− Ag+ Br−

Fig. 3.2 Frenkel defect in silver halides

Considering a diagrammatic representation of the oxidation process shown in Fig. 3.3 it is seen that either neutral atoms or ions *and* electrons must migrate in order for the reaction to proceed. In these cases the transport step of the reaction mechanism links the two phase boundary reactions as indicated. It will be noticed that there is an important distinction between scale growth by cation migration and scale growth by anion migration in that cation migration leads to scale formation at the scale–gas interface whereas anion migration leads to scale formation at the metal–scale interface.

M
MO
O2
M2+
2e−
M = M2+ + 2e−
M2+ + 2e− + ½O2 = MO
(a)
M
MO
O2
O2−
2e−
M + O2− = MO + 2e−
½O2 + 2e− = O2−
(b)

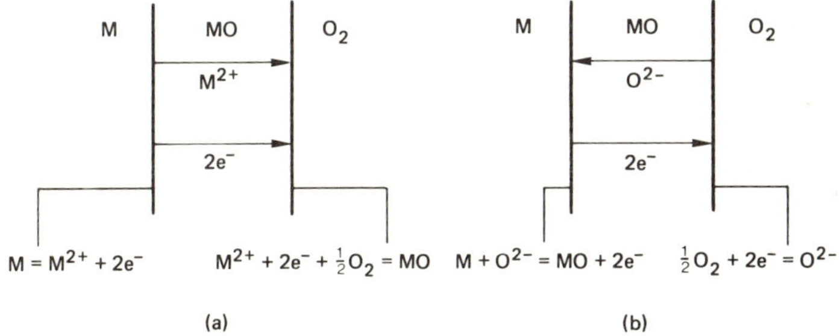

Fig. 3.3 Interfacial reactions and transport processes for high temperature oxidation mechanisms (a) cation mobile and (b) anion mobile

In order to explain simultaneous migration of ions and electrons it is necessary to assume that the oxides etc., that are formed during oxidation, are non-stoichiometric compounds.

Non-stoichiometric ionic compounds

By non-stoichiometry it is implied that the metal to non-metal atom ratio is not exactly that given by the chemical formula, even though the compound is electrically neutral. These can only be reconciled by assuming that either the anion or the cation exhibits variable valency on its sub-lattice. It is much more likely that the metal or cation shows variable valency.

Non-stoichiometric ionic compounds are classified as semiconductors and may show negative or positive behaviour.

N-type or negative semiconductors

The classification refers to the fact that electrical charge is transferred by negative carriers. This may arise, as shown below, by having either an excess of metal or a deficit of non-metal.

Metal excess

The chemical formula is given as $M_{1+\delta}O$ and the best known example is zinc oxide (ZnO). In order to allow extra metal in the compound, it is necessary to postulate the existence of interstitial cations with an equivalent number of electrons in the conduction band. The structure may be represented as shown in Fig. 3.4. Here, both Zn^+ and Zn^{2+} are represented as possible occupiers of interstitial sites. Cation conduction occurs over interstitial sites and electrical conductance occurs by virtue of having the 'excess' electrons excited into the conduction band. These, therefore, are called 'excess' or 'quasi-free' electrons.

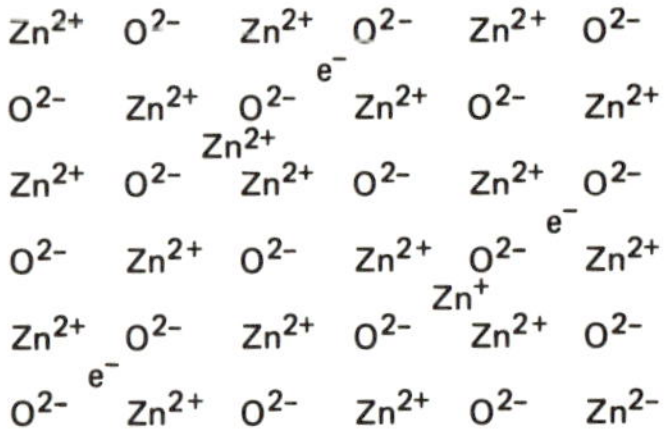

Zn^{2+} O^{2-} Zn^{2+} O^{2-} Zn^{2+} O^{2-}
e^-
O^{2-} Zn^{2+} O^{2-} Zn^{2+} O^{2-} Zn^{2+}
Zn^{2+}
Zn^{2+} O^{2-} Zn^{2+} O^{2-} Zn^{2+} O^{2-}
e^-
O^{2-} Zn^{2+} O^{2-} Zn^{2+} O^{2-} Zn^{2+}
Zn^+
Zn^{2+} O^{2-} Zn^{2+} O^{2-} Zn^{2+} O^{2-}
e^-
O^{2-} Zn^{2+} O^{2-} Zn^{2+} O^{2-} Zn^{2-}

Fig. 3.4 Interstitial cations and excess electrons in ZnO – an n-type metal excess semiconductor

The formation of this defect may be visualised, conveniently, as being formed from a perfect ZnO crystal by losing oxygen; the remaining unpartnered Zn^{2+} leaving the cation lattice and entering interstitial sites, and the two negative charges of the oxygen ion entering the conduction band. In this way one unit of ZnO crystal is destroyed and the formation of the defect may be represented as

$$ZnO = Zn_i^{\cdot\cdot} + 2e' + \tfrac{1}{2}O_2 \qquad (3.1)$$

for the formation of $Zn_i^{\cdot\cdot}$, doubly charged Zn interstitials

$$\text{or} \quad ZnO = Zn_i^{\cdot} + e' + \tfrac{1}{2}O_2 \qquad (3.2)$$

for the formation of $Zn_i^{\cdot}$, singly charged Zn interstitials.

These processes are shown diagrammatically in Fig. 3.5.

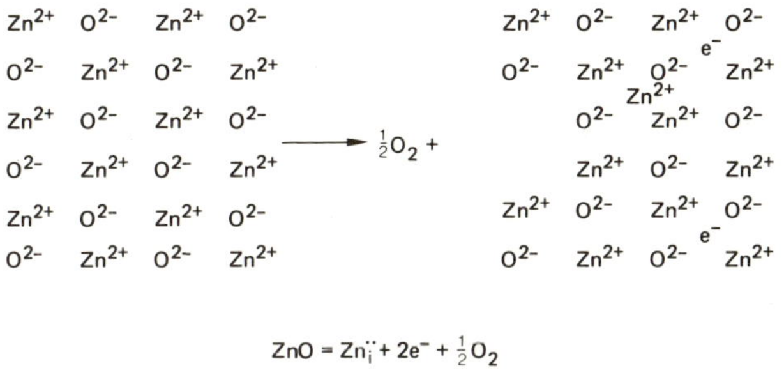

Zn^{2+} O^{2-} Zn^{2+} O^{2-} | Zn^{2+} O^{2-} Zn^{2+} O^{2-}
O^{2-} Zn^{2+} O^{2-} Zn^{2+} | e^-
Zn^{2+} O^{2-} Zn^{2+} O^{2-} | O^{2-} Zn^{2+} O^{2-} Zn^{2+}
O^{2-} Zn^{2+} O^{2-} Zn^{2+} → $\frac{1}{2}O_2$ + | Zn^{2+}
Zn^{2+} O^{2-} Zn^{2+} O^{2-} | O^{2-} Zn^{2+} O^{2-}
O^{2-} Zn^{2+} O^{2-} Zn^{2+} | Zn^{2+} O^{2-} Zn^{2+}
| Zn^{2+} O^{2-} Zn^{2+} O^{2-}
| e^-
| O^{2-} Zn^{2+} O^{2-} Zn^{2+}

$$ZnO = Zn_i^{\cdot\cdot} + 2e^- + \tfrac{1}{2}O_2$$

Fig. 3.5 Formation of metal excess ZnO with excess electrons and interstitial zinc, $Zn_i^{\cdot\cdot}$, ions from 'perfect' ZnO

The nomenclature used here is that of Kröger[1]

M_M – M atom on M site
X_X – X atom on X site
M_i – M atom on interstitial site
X_i – X atom on interstitial site
N_M – impurity N on M site
V_M – vacancy on M site
V_X – vacancy on X site
V_i – vacant interstitial
e' – electron in conduction band
$h^{\cdot}$ – electron hole in valence band

The charges on the defects are measured relative to the normal site occupation and are indicated by the superscripts ($^{\cdot}$) and ($'$) for positive and negative charges respectively, e.g. $Zn_i^{\cdot\cdot}$ represents a divalent zinc ion on an interstitial site and carries a charge of +2 relative to the normal, unoccupied, site.

The two equilibria shown above will yield to thermodynamic treatment giving the following for the equilibrium in equation 3.1

$$K_1 = a_{Zn_i^{\cdot\cdot}}\, a_{e'}^2\, p_{O_2}^{1/2} \tag{3.3}$$

or, since the defects are in very dilute solution, we may assume that they are in the range obeying Henry's law when the equilibrium may be written, in terms of concentrations $C_{Zn_i^{\cdot\cdot}}$ and $C_{e'}$, as

$$K_1' = C_{Zn_i^{\cdot\cdot}}\, C_{e'}^2\, p_{O_2}^{1/2} \tag{3.4}$$

If equation 3.1 represents the only mechanism by which defects are created in ZnO then it follows that

$$2C_{Zn_i^{\cdot\cdot}} = C_{e'} \tag{3.5}$$

Hence, putting equation 3.5 into equation 3.4

$$K_1' = 4C^3_{Zn_i^{\cdot\cdot}}\, p^{1/2}_{O_2}$$

or $$C_{Zn_i^{\cdot\cdot}} = (K_1'/4)^{1/3}\, p_{O_2}^{-1/6} = \text{const.}\ p_{O_2}^{-1/6} \tag{3.6}$$

i.e. $$C_{e'} \propto p_{O_2}^{-1/6} \tag{3.7}$$

Similarly applying the same analysis to reaction 3.2 the result is obtained that

$$C_{Zn_i^{\cdot}} = C_{e'} \propto p_{O_2}^{-1/4} \tag{3.8}$$

Clearly each mechanism predicts a different dependance of defect concentration on oxygen partial pressure. Since the electrical conductivity depends on the concentration of conduction band electrons, measurement of electrical conductivity as a function of oxygen partial pressure should serve to define the dependence law and indicate which defect is predominant. Initial experiments of this type carried out between 500 and 700 °C[3] indicated that the conductivity varied with oxygen partial pressure according to $p_{O_2}^{(-1/4.5\ -(-1/5.0))}$. This indicates that neither defect mechanism predominates, and the actual structure could involve both singly and doubly charged interstitial cations.

More recent work has confirmed that significant interstitial solution of zinc occurs in ZnO; the results of various studies being interpreted in terms of neutral zinc interstitial atoms[4] and singly charged interstitial ions[5,6], the latter being favoured presently[2], although it has also been suggested, on the basis of oxygen diffusion measurements, that oxygen vacancies are important above 1000 °C. Kofstad[2] concludes that no overall consistent interpretation can be made regarding ZnO.

Inconsistencies between different investigations concerning the establishment of defect structures can frequently result from impurities in the sample, especially in the case of compounds which show only slight deviations from stoichiometry. As the purity of available materials improves, it would be prudent to repeat such measurements.

Non-metal deficit

Alternatively, n-type behaviour can be caused by non-metal deficit as mentioned above. For oxides this may be visualised as the discharge and subsequent evaporation of an oxygen ion; the electrons enter the conduction band and a vacancy is created on the anion lattice. The process is shown in Fig. 3.6 and may be represented as

$$O_O = V_O^{\cdot\cdot} + 2e' + \tfrac{1}{2}O_2 \tag{3.9}$$

The vacant oxygen ion site surrounded by positive ions represents a site of

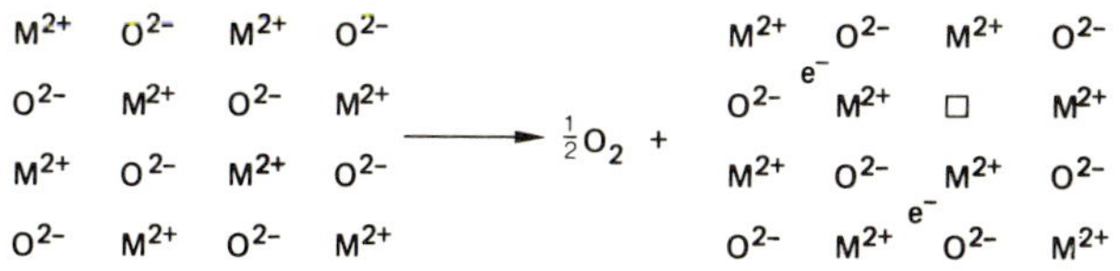

M^{2+} O^{2-} M^{2+} O^{2-} → $\frac{1}{2}O_2$ + M^{2+} O^{2-} M^{2+} O^{2-}

O^{2-} M^{2+} O^{2-} M^{2+} → O^{2-} e^- M^{2+} □ M^{2+}

M^{2+} O^{2-} M^{2+} O^{2-} → M^{2+} O^{2-} M^{2+} O^{2-}

O^{2-} M^{2+} O^{2-} M^{2+} → O^{2-} M^{2+} e^- O^{2-} M^{2+}

$$O_O = V_O^{\bullet\bullet} + 2e^- + \tfrac{1}{2}O_2$$

Fig. 3.6 Formation of oxygen deficit MO, with oxygen ion vacancies and excess electrons, from 'perfect' MO

high positive charge to which a free electron may be attracted. It is therefore possible for the vacancy and the free electrons to become associated according to

$$V_O^{\bullet\bullet} + e' = V_O^{\bullet} \tag{3.10}$$

and

$$V_O^{\bullet} + e' = V_O \tag{3.11}$$

Therefore, we have the possibility that doubly and singly charged vacancies may exist as well as neutral vacancies.

The question of the charges on defects is not always obvious and must be considered from the aspect of the perfect lattice. If a site which is normally occupied by an anion (O^{2-}) is vacant, then the lattice lacks two negative charges at that site compared with the normal lattice. The vacant anion site is therefore regarded as having two positive charges with respect to the normal lattice. Of course, such a site would represent an energy trough for electrons and so may capture conduction band electrons, thus reducing the relative positive charge in the vacant site as indicated in equations 3.10 and 3.11 above.

By noting the relationships between the concentrations of anion vacancies and conduction band electrons, the dependence of electrical conductivity on oxygen partial pressure can be derived as before. These are, for the doubly charged vacancy

$$\chi \propto C_{e'} \propto p_{O_2}^{-1/6} \tag{3.12}$$

for the singly charged vacancy

$$\chi \propto C_{e'} \propto p_{O_2}^{-1/4} \tag{3.13}$$

and for the neutral vacancy

$$\chi \propto C_{e'} \quad (\text{independent of } p_{O_2}) \tag{3.14}$$

Notice that in equations 3.7, 3.8, 3.12, and 3.13 the power of the oxygen partial pressure is negative. This is the case for all n-type semiconductors.

P-type or positive semiconductors

In this case charge is transferred by positive carriers. This may arise from either a deficit of metal or an excess of non-metal (intrinsic semiconduction).

Metal deficit

P-type semiconduction arises from the formation of vacancies on the cation lattice together with electron holes giving rise to conduction. The formula may be written $M_{1-\delta}O$. The value of δ can vary widely; from 0.05 in the case of $Fe_{0.95}O$ (wustite), 0.001 for NiO, to very small deviations in the case of Cr_2O_3 and Al_2O_3.

The possibility of forming electron holes lies in the ability of many metal ions, especially of the ions of transition metals, to exist in several valence states. The closer together the different valence states are in terms of energy of ionisation, the more easily can cation vacancy formation be induced by the metal deficit p-type mechanism.

The typical structure of this class of oxides is represented in Fig. 3.7 taking

Ni^{2+}	O^{2-}	Ni^{2+}	O^{2-}	Ni^{3+}	O^{2-}
O^{2-}	Ni^{2+}	O^{2-}	Ni^{2+}	O^{2-}	Ni^{2+}
Ni^{2+}	O^{2-}	Ni^{3+}	O^{2-}	□	O^{2-}
O^{2-}	Ni^{2+}	O^{2-}	Ni^{3+}	O^{2-}	Ni^{2+}
Ni^{2+}	O^{2-}	Ni^{2+}	O^{2-}	Ni^{2+}	O^{2-}
O^{2-}	□	O^{2-}	Ni^{2+}	O^{2-}	Ni^{2+}
Ni^{2+}	O^{2-}	Ni^{3+}	O^{2-}	Ni^{2+}	O^{2-}

Fig. 3.7 Typical p-type metal deficit semiconductor NiO, with cation vacancies and electron holes

NiO as an example. By virtue of the energetically close valence states of the cation it is relatively easy for an electron to transfer from a Ni^{2+} to a Ni^{3+} thus reversing the charges on the two ions. The site Ni^{3+} is thus seen to offer a low energy position for an electron and is called an 'electron hole'.

The formation of this defect structure can easily be visualised if one considers the interaction of the NiO lattice with oxygen shown in Fig. 3.8.

In step (b) the oxygen chemisorbs by attracting an electron from a Ni^{2+} site thus forming a Ni^{3+} or hole. In step (c) the chemisorbed oxygen is fully ionised forming another hole and a Ni^{2+} ion enters the surface to partner the O^{2-} thus forming a vacancy in the cation sub-lattice. Note that this process also forms an extra unit of NiO on the surface of the oxide which should reflect in density changes if sufficiently sensitive measurements were made.

The process can be represented by the equation

$$\tfrac{1}{2}O_2 = O_O + 2h^{\cdot} + V''_{Ni} \qquad (3.15)$$

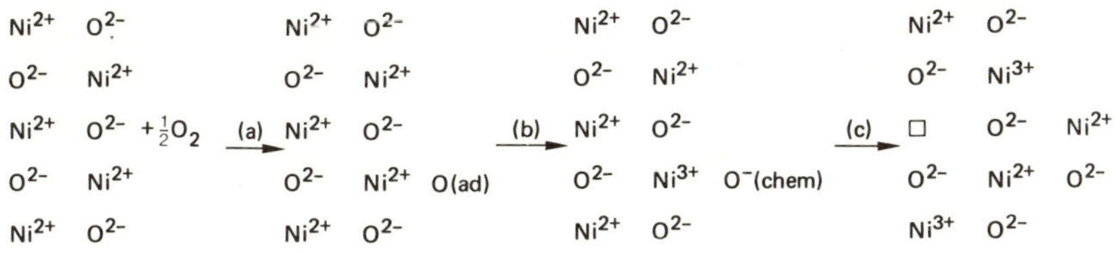

Ni^{2+} O^{2-}	Ni^{2+} O^{2-}	Ni^{2+} O^{2-}	Ni^{2+} O^{2-}
O^{2-} Ni^{2+}	O^{2-} Ni^{2+}	O^{2-} Ni^{2+}	O^{2-} Ni^{3+}
Ni^{2+} O^{2-} $+\frac{1}{2}O_2$ $\xrightarrow{(a)}$	Ni^{2+} O^{2-} $\xrightarrow{(b)}$	Ni^{2+} O^{2-} $\xrightarrow{(c)}$	$\square$ O^{2-} Ni^{2+}
O^{2-} Ni^{2+}	O^{2-} Ni^{2+} O(ad)	O^{2-} Ni^{3+} O^-(chem)	O^{2-} Ni^{2+} O^{2-}
Ni^{2+} O^{2-}	Ni^{2+} O^{2-}	Ni^{2+} O^{2-}	Ni^{3+} O^{2-}

(a) Adsorption: $\frac{1}{2}O_2$ (g) = O (ad)

(b) Chemisorption: O(ad) = O^-(chem) + $h^{\cdot}$

(c) Ionisation: O^-(chem) = $O_O + V''_{Ni} + h^{\cdot}$

Overall reaction: $\frac{1}{2}O_2 = O_O + V''_{Ni} + 2h^{\cdot}$

Fig. 3.8 Formation of metal deficit p-type semiconductor, with cation vacancies and electron holes by incorporation of oxygen into the 'perfect' lattice

the equilibrium constant for which may be written as below, assuming that equation 3.15 represents the only mechanism by which the defects form and that these obey Henry's law.

$$C^2_{h^{\cdot}} \cdot C_{V''_{Ni}} = K p^{1/2}_{O_2} \quad (3.16)$$

By stoichiometry $C_{h^{\cdot}} = 2C_{V''_{Ni}}$ for electrical neutrality, and thus

$$C_{h^{\cdot}} = \text{const.}\, p^{1/6}_{O_2} \quad (3.17)$$

The electrical conductivity is expected to vary proportionally to the electron hole concentration and so with the sixth root of the oxygen partial pressure. A further possibility, however, is that a cation vacancy may bond with an electron hole, i.e. in NiO a Ni^{3+} may be permanently attached to the Ni vacancy. In this case, the formation of vacancies and electron holes, on oxidising, may be represented as

$$\tfrac{1}{2}O_2 = O_O + h^{\cdot} + V'_{Ni} \quad (3.18)$$

in this case we find

$$C_{h^{\cdot}} = \text{const.}\, p^{1/4}_{O_2} \quad (3.19)$$

Experimental results on the variation of electrical conductivity with oxygen partial pressure have confirmed that both types of defect are found in NiO[2,3,8–11].

Equations 3.15 and 3.18 are clearly linked by the equation

$$V''_{Ni} + h^{\cdot} = V'_{Ni} \quad (3.20)$$

Since the concentration of doubly charged vacancies will increase according to $p^{1/6}_{O_2}$ and that of singly charged vacancies according to $p^{1/4}_{O_2}$, it is to be expected that doubly charged vacancies will predominate at low oxygen partial pressures and that singly charged vacancies will predominate at high oxygen partial pressures. Thus, the electrical conductivity is expected to vary with oxygen partial pressure according to a one sixth power at low partial pressures and according to a one fourth power at high partial pressures, corresponding to equations 3.17 and 3.19 respectively. Some evidence exists for this behaviour[12,13,23].

It should be noticed that in equations 3.17 and 3.19 the power of the oxygen partial pressure is positive. This is so for all p-type semiconducting oxides.

Intrinsic semiconductors

There is another type of oxide which, although having quite close stoichiometry, shows relatively high electrical conductivity which is independent of the oxygen partial pressure. Such behaviour is typical of so called 'intrinsic', or 'transitional', semiconduction when the concentration of electronic defects far exceeds that of ionic defects and the equilibrium of the electronic defects may be represented by the excitation of an electron from the valence state to the conduction band, producing a 'quasi-free' electron and an electron hole according

$$\text{Null} = h^{\cdot} + e' \qquad (3.21)$$

The oxide CuO behaves in this way at 1000 °C. Very few oxides belong to this class which is not of great importance for an introductory study of the subject of high temperature oxidation of metals.

Rates of oxidation

It has already been seen that, for an oxidation process to proceed, under conditions where the two reactants are separated by the reaction product, it is necessary to postulate that ionic and electronic transport processes through the oxide are accompanied by ionising phase boundary reactions and formation of new oxide at a site whose position depends upon whether cations or anions are transported through the oxide layer.

Taking this very simple model, and making still more simplifying assumptions, Wagner[14] was able to develop his celebrated theory of the high temperature oxidation of metals. In fact, the theory describes the oxidation behaviour only for the case where diffusion of ions is rate determining and under highly idealised conditions. Before proceeding with the derivation of Wagner's expressions, a simplified treatment will be presented which may serve to emphasise the important features of diffusion controlled oxidation.

Simplified treatment of diffusion controlled oxidation

Assuming that ionic transport across the growing oxide layer controls the rate of scaling and that thermodynamic equilibrium is established at each interface, the process can be analysed as follows. The outward cation flux, $j_{M^{2+}}$, is equal and opposite to the inward flux of cation defects (here taken to be vacancies). This model is shown in Fig. 3.9.

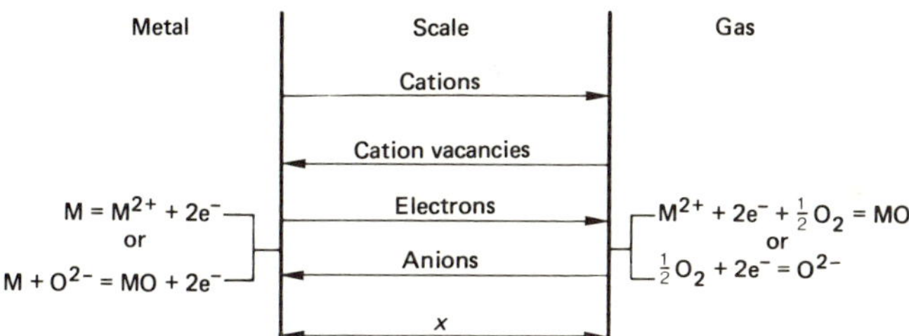

Fig. 3.9 Simplified model for diffusion controlled oxidation

Thus

$$j_{M^{2+}} = -j_{V_M} = D_{V_M}\frac{C''_{V_M} - C'_{V_M}}{x} \tag{3.22}$$

where x is the oxide thickness, D_{V_M} is the diffusion coefficient for cation vacancies, and C''_{V_M} and C'_{V_M} are the vacancy concentrations at the scale–gas and scale–metal interfaces respectively.

Since there is thermodynamic equilibrium at each interface, the value $(C''_{V_M} - C'_{V_M})$ is constant and we have

$$j_{V_M} = \text{const.}\,\frac{dx}{dt} = D_{V_M}\frac{C''_{V_M} - C'_{V_M}}{x} \tag{3.23}$$

$$\text{i.e.}\ \frac{dx}{dt} = \frac{k'}{x} \quad \text{where } k' = D_{V_M}(C''_{V_M} - C'_{V_M})/\text{const.} \tag{3.24}$$

Integrating and noting that $x = 0$ at $t = 0$

$$x^2 = 2k't \tag{3.25}$$

which is the common parabolic rate law.

Furthermore, since it has been shown that the cation vacancy concentration is related to the oxygen partial pressure by

$$C_{V_M} = \text{const.}\ p_{O_2}^{1/n}$$

the variation of the parabolic rate constant with oxygen partial pressure can be predicted i.e. $k' \propto [(p''_{O_2})^{1/n} - (p'_{O_2})^{1/n}]$ (3.26)

Since p'_{O_2} is usually negligible compared with p''_{O_2} we have

$$k' \propto (p''_{O_2})^{1/n} \quad (3.27)$$

Wagner theory of oxidation[14]

Figure 3.10 gives a summary of the conditions, later stated, for which the theory is valid.

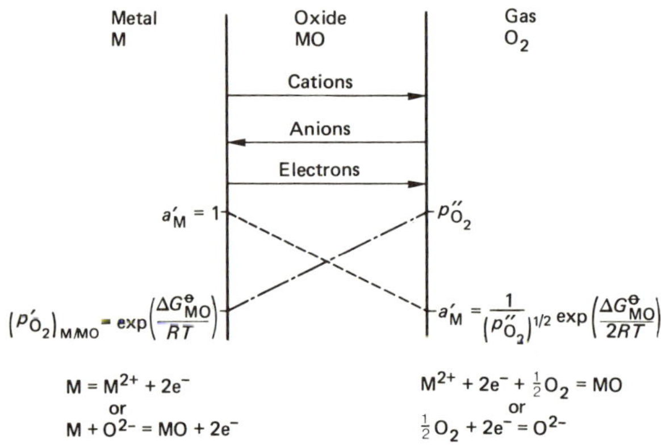

Fig. 3.10 Diagram of scale formation according to Wagner's model

Assumptions

(1) The oxide layer is a compact, perfectly adherent scale.
(2) Migration of ions or electrons across the scale is the rate controlling process.
(3) Thermodynamic equilibrium is established at both the metal–scale and scale–gas interfaces.
(4) The oxide scale shows only small deviations from stoichiometry.
(5) Thermodynamic equilibrium is established locally throughout the scale.
(6) The scale is thick compared with the distances over which space charge effects (electrical double layer) occur.
(7) Oxygen solubility in the metal may be neglected.

Since thermodynamic equilibrium is assumed to be established at the metal–scale and scale–gas interfaces, it follows that activity gradients of both metal and non-metal (oxygen, sulphur, etc.) are established across the scale. Consequently, metal ions and oxygen ions will tend to migrate across the scale in opposite directions. Because the ions are charged, this migration will cause an electric field to be set up across the scale resulting in consequent transport of electrons across the scale from metal to atmosphere. The relative migration rates of cations, anions, and electrons are therefore balanced such that no net charge transfer occurs across the oxide layer as a result of ionic migration.

Being charged particles, ions will respond to both chemical and electrical potential gradients which together provide the net driving force for ionic migration.

A particle, i, carrying a charge, Z_i, in a position where the chemical potential gradient is $\partial\mu_i/\partial x$ and the electric potential gradient is $\partial\phi/\partial x$ is acted on by a force given by $\partial\mu_i/\partial x + Z_iF(\partial\phi/\partial x)$ joules $mol^{-1}\,cm^{-1}$

$$\text{or}\quad 1/N_A\left(\frac{\partial\mu_i}{\partial x} + Z_iF\frac{\partial\phi}{\partial x}\right) \text{ joules particle}^{-1}\,\text{cm}^{-1} \tag{3.28}$$

where N_A is Avogadro's number and F is the Faraday constant in coulomb mol^{-1}.

Moving through an ionic lattice under a constant force, an ion rapidly assumes a constant velocity known as the terminal drift velocity. The value of the terminal drift velocity when the ion is acted on by unit force is known as the mobility, B_i, of the species. Therefore, the flux of i, acted on by the force given in equation 3.28, is

$$j_i = c_iB_i\,1/N_A\left(\frac{\partial\mu_i}{\partial x} + Z_iF\frac{\partial\phi}{\partial x}\right) \text{ particles cm}^{-2}\,\text{s}^{-1} \tag{3.29}$$

where c_i is the concentration of i in particles cm^{-3}.

Alternatively

$$j_i = -\frac{c_iB_i}{N_A^2}\left(\frac{\partial\mu_i}{\partial x} + Z_iF\frac{\partial\phi}{\partial x}\right) \text{ mols cm}^{-2}\,\text{s}^{-1} \tag{3.30}$$

The mobility and partial conductivity of species i are related by

$$c_iB_i = \frac{\sigma_i}{Z_i^2\,e^2} \tag{3.31}$$

where σ_i is the partial electrical conductivity of i and e is the electronic charge, thus

$$j_i = -\frac{\sigma_i}{Z_i^2F^2}\left(\frac{\partial\mu_i}{\partial x} + Z_iF\frac{\partial\phi}{\partial x}\right) \tag{3.32}$$

Equation 3.32 can be used to describe the flux of cations, anions, or electrons through the oxide layer. Due to their different mobilities, different species would tend to move at different rates, however, this would set up electric fields tending to oppose this independence. In fact, the three species migrate at rates that are defined by the necessity of maintaining electroneutrality throughout the scale, i.e. such that there is no net charge across the oxide scale. This condition is usually achieved due to the very high mobility of electrons or electronic defects.

The original treatment by Wagner involved cations, anions, and electrons. However, the majority of oxides and sulphides show high electronic mobility, and the mobilities of the cation and anion species usually differ by several orders of magnitude. It is, therefore, possible to ignore the migration of the slower moving ionic species and thus simplify the treatment somewhat.

The most frequently met case involves oxides and sulphides in which cations and electrons are the mobile species. Writing Z_c and Z_e for the charges on cations and electrons respectively, and using equation 3.32, we have for the respective fluxes

$$j_c = -\frac{\sigma_c}{Z_c^2 F^2}\left(\frac{\partial \mu_c}{\partial x} + Z_c F \frac{\partial \phi}{\partial x}\right) \tag{3.33}$$

and

$$j_e = -\frac{\sigma_e}{Z_e^2 F^2}\left(\frac{\partial \mu_e}{\partial x} + Z_e F \frac{\partial \phi}{\partial x}\right) \tag{3.34}$$

The condition for electrical neutrality is

$$Z_c j_c + Z_e j_e = 0 \tag{3.35}$$

and, using equations 3.33, 3.34, and 3.35, it is now possible to eliminate $\partial\phi/\partial x$ as

$$\frac{\partial \phi}{\partial x} = -\frac{1}{F(\sigma_c + \sigma_e)}\left[\frac{\sigma_c}{Z_c}\frac{\partial \mu_c}{\partial x} + \frac{\sigma_e}{Z_e}\frac{\partial \mu_e}{\partial x}\right] \tag{3.36}$$

Substituting equation 3.36 in equation 3.33 we have for the flux of cations

$$j_c = -\frac{\sigma_c \sigma_e}{Z_c^2 F^2(\sigma_c + \sigma_e)}\left[\frac{\partial \mu_c}{\partial x} - \frac{Z_c}{Z_e}\frac{\partial \mu_e}{\partial x}\right] \tag{3.37}$$

Since the electronic charge $Z_e = -1$, equation 3.37 becomes

$$j_c = -\frac{\sigma_c \sigma_e}{Z_c^2 F^2(\sigma_c + \sigma_e)}\left[\frac{\partial \mu_c}{\partial x} + Z_c \frac{\partial \mu_e}{\partial x}\right] \tag{3.38}$$

The ionisation of a metal M is represented by

$$M = M^{Z_c} + Z_c e$$

and it follows that, at equilibrium

$$\mu_M = \mu_c + Z_c \mu_e \tag{3.39}$$

Therefore, from equations 3.38 and 3.39

$$j_c = -\frac{\sigma_c\sigma_e}{Z_c^2F^2(\sigma_c + \sigma_e)}\frac{\partial\mu_M}{\partial x} \tag{3.40}$$

Equation 3.40 is the expression for the cationic flux at any place in the scale, where σ_c, σ_e, and $\partial\mu_M/\partial x$ are the instantaneous values at that place. Since all of these values may change with the position in the scale, it is necessary to integrate equation 3.40 in order to define j_c in terms of the scale thickness, x, and the measurable metal chemical potentials at the metal–scale, μ'_M, and scale–gas, μ''_M, interfaces, i.e.

$$j_c = -\frac{1}{Z_c^2F^2x}\int_{\mu'_M}^{\mu''_M}\frac{\sigma_c\sigma_e}{\sigma_c + \sigma_e}\,\mathrm{d}\mu_M$$

or

$$j_c = \frac{1}{Z_c^2F^2x}\int_{\mu''_M}^{\mu'_M}\frac{\sigma_c\sigma_e}{\sigma_c + \sigma_e}\,\mathrm{d}\mu_M \text{ mol cm}^{-2}\text{ s}^{-1} \tag{3.41}$$

If the concentration of metal in the oxide scale is C_M mol cm^{-3} then the flux may also be expressed by

$$j_c = C_M\frac{\mathrm{d}x}{\mathrm{d}t} \tag{3.42}$$

where x is the oxide scale thickness.

The parabolic rate law is usually expressed as

$$\frac{\mathrm{d}x}{\mathrm{d}t} = \frac{k'}{x} \tag{3.43}$$

where k' is the parabolic rate constant in cm^2 s^{-1}.

Comparing equations 3.41, 3.42, and 3.43, the parabolic rate constant is expressed by

$$k' = \frac{1}{Z_c^2F^2C_M}\int_{\mu''_M}^{\mu'_M}\frac{\sigma_c\sigma_e}{\sigma_c + \sigma_e}\,\mathrm{d}\mu_M \tag{3.44}$$

A similar treatment for the case where anions are more mobile than cations, i.e. the migration of cations may be neglected, yields

$$k' = \frac{1}{Z_a^2F^2C_X}\int_{\mu'_X}^{\mu''_X}\frac{\sigma_a\sigma_e}{\sigma_a + \sigma_e}\,\mathrm{d}\mu_X \tag{3.45}$$

where σ_a is the partial electrical conductivity of the anions and X is the non-metal, oxygen or sulphur.

In general, it is found that the transport number of electrons, or electronic defects, is close to unity, compared with which the transport numbers of cations or anions are negligibly small. In these cases equations 3.44 and 3.45 reduce to

$$k' = \frac{1}{Z_c^2 F^2 C_M} \int_{\mu_M''}^{\mu_M'} \sigma_c \, d\mu_M \text{ cm}^2 \text{ s}^{-1} \tag{3.46}$$

and

$$k' = \frac{1}{Z_a^2 F^2 C_X} \int_{\mu_X'}^{\mu_X''} \sigma_a \, d\mu_X \text{ cm}^2 \text{ s}^{-1} \tag{3.47}$$

respectively.

The mobility, B_i, and diffusion coefficient, D_i, of a species are related by the Nernst–Einstein equation

$$D_i = B_i k T \tag{3.48}$$

where k is Boltzmann's constant and T is absolute temperature. Introducing equation 3.31

$$c_i B_i = \frac{\sigma_i}{Z_i^2 e^2} \tag{3.31}$$

we have

$$D_i = \frac{k T \sigma_i}{Z_i^2 e^2 c_i} \tag{3.49}$$

where c_i is the concentration in particles cm^{-3}. Thus

$$D_i = \frac{RT}{Z_i^2 F^2 C_i} \tag{3.50}$$

where C_i is the concentration in mols cm^{-3}.

Consequently, using equations 3.46, 3.48, and 3.49, we have

$$k' = \frac{1}{RT} \int_{\mu_M''}^{\mu_M'} D_M \, d\mu_M \text{ cm}^2 \text{ s}^{-1} \tag{3.51}$$

and

$$k' = \frac{1}{RT} \int_{\mu_X'}^{\mu_X''} D_X \, d\mu_X \text{ cm}^2 \text{ s}^{-1} \tag{3.52}$$

for the parabolic rate constant, where D_M and D_X are the diffusion coefficients for metal, M, and non-metal, X, through the scale respectively.

Equations 3.51 and 3.52 are written in terms of variables that can be measured relatively easily, although it is assumed that the diffusion coefficient is a function of the chemical potential of the species involved.

Thus, in order to be able to calculate values of the parabolic rate constant, the relevant diffusion coefficient must be known as a function of the chemical potential of the mobile species. Such data are frequently not available or are incomplete. Furthermore, it is usually easier to measure the parabolic rate constant directly than to carry out experiments to measure the diffusion data. Thus, the real value of Wagner's analysis lies in providing a complete mechanistic understanding of the process of high temperature oxidation under the conditions set out.

The predictions of Wagner's theory for n-type and p-type oxides in which cations are the mobile species should now be examined. The first class is represented by ZnO and the oxidation of zinc.

The defect structure of zinc oxide involves interstitial zinc ions and excess or quasi-free electrons according to

$$ZnO = Zn_i^{\cdot\cdot} + 2e' + \tfrac{1}{2}O_2 \tag{3.1}$$

and

$$ZnO = Zn_i^{\cdot} + e' + \tfrac{1}{2}O_2 \tag{3.2}$$

The partial conductivity of interstitial zinc ions is expected to vary with their concentration. As shown before it is expected that this will vary according to the oxygen partial pressure

$$\sigma_{Zn_i^{\cdot\cdot}} \propto C_{Zn_i^{\cdot\cdot}} \propto p_{O_2}^{-1/6} \tag{3.53}$$

and

$$C_{Zn_i^{\cdot}} \propto p_{O_2}^{-1/4} \tag{3.54}$$

i.e.

$$\sigma_{Zn_i^{\cdot\cdot}} = \text{const.}\, p_{O_2}^{-1/6} \tag{3.55}$$

and

$$\sigma_{Zn_i^{\cdot}} = \text{const}'\, p_{O_2}^{-1/4} \tag{3.56}$$

Since ZnO shows only small deviations from stoichiometry

$$\mu_{Zn} + \mu_O = \mu_{ZnO} \approx \text{constant} \tag{3.57}$$

also

$$\mu_O = \tfrac{1}{2}\mu_{O_2}^{\ominus} + \tfrac{1}{2}RT\ln p_{O_2} \tag{3.58}$$

Therefore

$$0 = d\mu_{Zn} + \tfrac{1}{2}RT\,d\ln p_{O_2}$$

or

$$d\mu_{Zn} = -\tfrac{1}{2}RT\,d\ln p_{O_2} \tag{3.59}$$

Using equations 3.46, 3.55, and 3.58 we obtain

$$k_{ZnO} = -\text{const.} \int_{p''_{O_2}}^{p'_{O_2}} p_{O_2}^{-1/6} \, d\ln p_{O_2} = -\text{const.} \int_{p''_{O_2}}^{p'_{O_2}} p_{O_2}^{-7/6} \, dp_{O_2}$$

i.e. integrating

$$k_{ZnO} = \text{const.}' \left[\left(\frac{1}{p'_{O_2}} \right)^{1/6} - \left(\frac{1}{p''_{O_2}} \right)^{1/6} \right] \tag{3.60}$$

Alternatively, using equations 3.46 and 3.56

$$k_{ZnO} = \text{const.}'' \left[\left(\frac{1}{p'_{O_2}} \right)^{1/4} - \left(\frac{1}{p''_{O_2}} \right)^{1/4} \right] \tag{3.61}$$

Generally p''_{O_2} is much greater than p'_{O_2} and so the parabolic rate constant for zinc is expected to be practically independent of the externally applied oxygen partial pressure, regardless of the relative abundances of $Zn_i^{\cdot\cdot}$ and $Zn_i^{\cdot}$.

A similar treatment can be applied to the oxidation of cobalt to CoO. CoO is a metal deficit p-type semiconductor forming cation vacancies and electron holes[15,16] according to

$$\tfrac{1}{2}O_2 = O_O + V'_{Co} + h^{\cdot} \tag{3.62}$$

$$K = C_{V'_{Co}} C_{h^{\cdot}} p_{O_2}^{-1/2}$$

Excluding defects caused by impurities and any contribution from intrinsic electronic defects we see from equation 3.62 that, in order to satisfy both stoichiometry and electroneutrality, we have

$$C_{V'_{Co}} = C_{h^{\cdot}} \propto p_{O_2}^{1/4} \tag{3.63}$$

and consequently the cationic partial conductivity is given by

$$\sigma_{Co} \propto p_{O_2}^{1/4} \tag{3.64}$$

Thus, substituting equation 3.64 into equation 3.47, we obtain for the parabolic scaling constant (for Co being oxidised to CoO)

$$k'_{Co} \propto \int_{p'_{O_2}}^{p''_{O_2}} p_{O_2}^{1/4} \, d\ln p_{O_2} \tag{3.65}$$

Alternatively

$$k'_{Co} \propto \int_{p'_{O_2}}^{p''_{O_2}} p_{O_2}^{-3/4} \, dp_{O_2} \tag{3.65a}$$

and integrating

$$k'_{Co} \propto [(p''_{O_2})^{1/4} - (p'_{O_2})^{1/4}] \tag{3.66}$$

In equation 3.66 the value of p'_{O_2} for the equilibrium between Co and CoO can be neglected compared with the value of p''_{O_2} in the atmosphere. This is valid because diffusion control of the oxidation reaction rate is achieved only under conditions where p''_{O_2} is sufficiently high to avoid control of the reaction rate by surface reactions or transport through the gas phase.

Under this assumption, equation 3.66 simplifies to

$$k'_{Co} \propto (p''_{O_2})^{1/4} \tag{3.67}$$

The existence of doubly charged cation vacancies might also be postulated according to

$$\tfrac{1}{2}O_2 = O_O + V''_{Co} + 2h^{\cdot} \tag{3.68}$$

for which

$$K = C_{V''_{Co}}\, C^2_{h^{\cdot}}\, p_{O_2}^{-1/2}$$

If these defects were predominant then, following the same argument, the parabolic rate constant for oxidation of cobalt would be expected to vary with oxygen partial pressure as

$$k'_{Co} \propto (p''_{O_2})^{1/6} \tag{3.69}$$

Measurements of the electrical conductivity of CoO as a function of oxygen partial pressure by Fisher and Tannhauser[15] show that the doubly charged vacancy model according to equation 3.68 is obeyed up to values of oxygen partial pressure in the region of 10^{-5} atm. Above this value the singly charged model according to equation 3.62 was obeyed. Since the oxygen partial pressure at which the change occurs is so low, it may be expected that the CoO scale, growing on cobalt, would contain defects according to equation 3.62, predominantly, except for a very thin layer immediately in contact with the metal. The effect of this would be negligible.

Thus, if cobalt is oxidised in atmospheres where the oxygen partial pressure is substantially greater than 10^{-5} atm, the rate constant is expected to vary with $p''^{1/4}_{O_2}$. (Note that, unless $p_{O_2} \gg 10^{-5}$ atm, parabolic scaling will not occur.)

The oxidation of cobalt has been investigated[17,18] over a range of oxygen partial pressures and temperatures. A plot of log k''_{Co} versus log p_{O_2} resulted in a series of lines of slope about 1/3.1 compared with the 1/4 expected. These results are shown in Fig. 3.11 and can be interpreted by the presence, in the CoO, also of neutral cation vacancies. Neutral cation vacancies form according to

$$\tfrac{1}{2}O_2 = O_O + V_{Co} \tag{3.70}$$

It is important to realise that, since their formation does not involve electrical defects, the presence of neutral vacancies will not be detected by electrical conductivity measurements made as a function of oxygen partial pressure. However, neutral vacancies can, and do, affect the diffusion of cations and, if equation 3.70 represents the majority of defect formation in

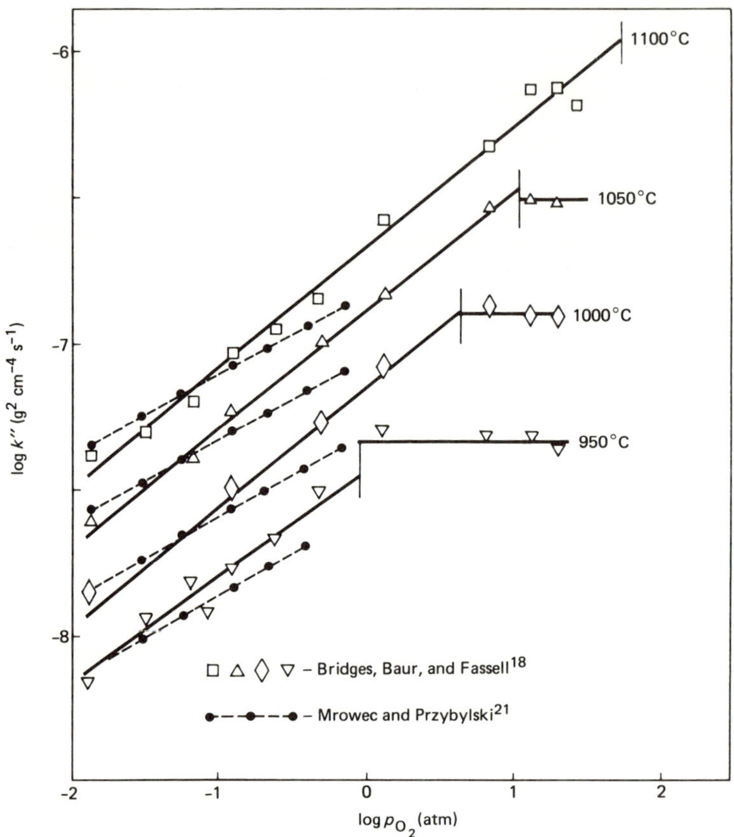

Fig. 3.11 Variation of the parabolic rate constant with oxygen partial pressure and temperature for the oxidation of cobalt, showing the results of Bridges, Baur, and Fassell[18] and Mrowec and Przybylski[21]

CoO, it would be expected that the diffusion coefficient of cobalt and the parabolic rate constant would vary as

$$D_{\mathrm{Co}} \propto p_{\mathrm{O}_2}^{1/2} \tag{3.71a}$$

and

$$k_{\mathrm{Co}} \propto p_{\mathrm{O}_2}^{1/2} \tag{3.71b}$$

The results shown in Fig. 3.11, which give the exponent of 1/3.1, confirm that in CoO singly charged and neutral vacancies are the important species.

Also to be seen from Fig. 3.11 is the fact that, as soon as the oxygen potential of the atmosphere exceeds that at which Co_3O_4 can form, the parabolic rate constant becomes insensitive to the external oxygen partial pressure.

As a further example of how the presence of several defect types can influence the results, let us consider the oxidation of copper to Cu_2O. Cu_2O is a metal deficit p-type semiconductor forming cation vacancies and electron holes.

The formation of these defects may be represented by the equation

$$\tfrac{1}{2}O_2 = O_O + 2V'_{Cu} + 2h^{\cdot} \tag{3.72}$$

However it is also possible for the vacancies and electron holes to associate, producing uncharged copper vacancies, according to

$$V'_{Cu} + h^{\cdot} = V_{Cu} \tag{3.73}$$

Alternatively, the formation of neutral cation vacancies may be represented by combining equations 3.72 and 3.73 as

$$\tfrac{1}{2}O_2 = O_O + 2V_{Cu} \tag{3.74}$$

Following the same derivation as in the case of CoO it is seen that, if equation 3.72 represents the defect structure of Cu_2O, then the parabolic rate constant for the oxidation of copper to Cu_2O should follow the relationship

$$k''_{Cu} \propto p_{O_2}^{1/8} \tag{3.75}$$

Alternatively, if the defect structure is represented by equation 3.74, i.e. neutral cation vacancies predominate, then the relationship for the parabolic rate constant should be

$$k''_{Cu} \propto p_{O_2}^{1/4} \tag{3.76}$$

Mrowec and Stoklosa[19] and Mrowec *et al.*[20] have investigated the oxidation of Cu to Cu_2O over a range of temperatures and oxygen partial pressures and their results are shown in Fig. 3.12. From this it is clear that the exponent is very close to 1/4, indicating that neutral cation vacancies are the predominant defect species in Cu_2O.

A more complete test of the Wagner mechanism and treatment has been carried out by Fueki and Wagner[23] and by Mrowec and Przybylski[21,22] who derived values for the diffusion coefficient of nickel in NiO and cobalt in CoO from measurements of the parabolic oxidation rate constant. Since the parabolic rate constant can be expressed in the form of equation 3.51

$$k' = \frac{1}{RT}\int_{\mu''_M}^{\mu'_M} D_M \, d\mu_M \tag{3.51}$$

it should be possible to calculate k' from knowledge of the diffusion coefficient D_M as a function of the chemical potential of M, μ_M. It is, however, easier to measure the oxygen partial pressure than the metal chemical potential and equation 3.51 can be rearranged accordingly.

For example, if the metal oxide is formed according to the general equation

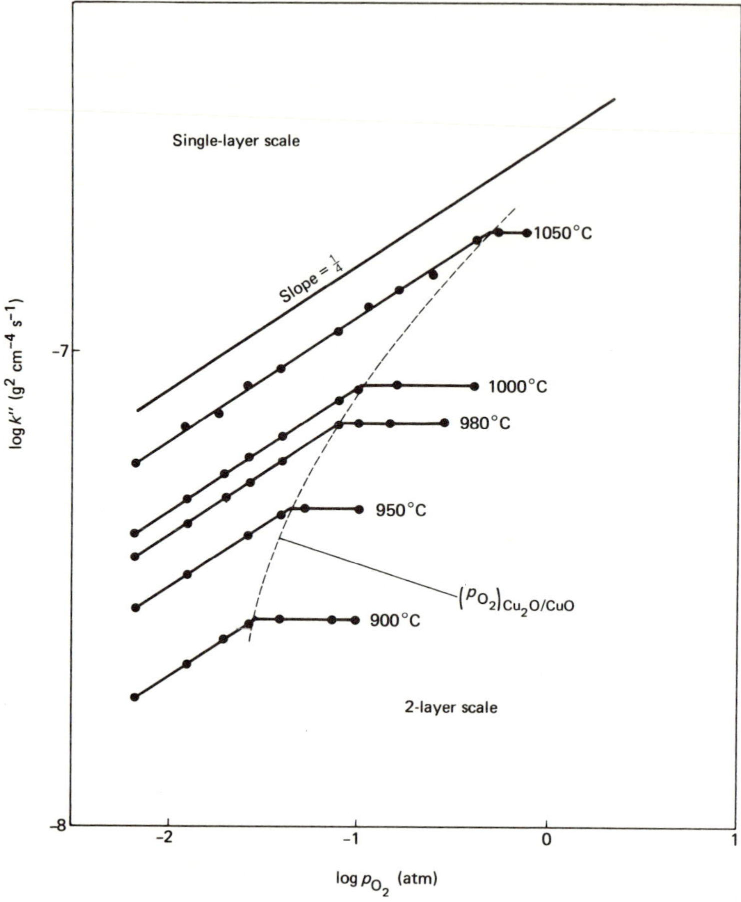

Fig. 3.12 Variation of the parabolic rate constant with oxygen partial pressure and temperature for the oxidation of copper, following Mrowec and Stoklosa[19] and Mrowec *et al.*[20]

$$2M + \frac{Z_c}{2} O_2 = M_2O_{Z_c} \tag{3.77}$$

and assuming that deviations from stoichiometry are small, such that the chemical potential of $M_2O_{Z_c}$ is constant, we have

$$2\mu_M + Z_c\mu_O = \mu_{M_2O_{Z_c}} = \text{const.} \tag{3.78}$$

from which, since $\mu_O = \frac{1}{2}(\mu^{\ominus}_{O_2} + RT\ln p_{O_2})$, we obtain

$$d\mu_M = -\frac{Z_c}{4} RT\, d\ln p_{O_2} \tag{3.79}$$

Substituting equation 3.73 in equation 3.51 the result is

$$k' = \frac{Z_c}{4}\int_{p'_{O_2}}^{p''_{O_2}} D_M \, d \ln p_{O_2} \tag{3.80}$$

The difficulty in using equation 3.80 to calculate values of k' lies in having to know how the value of D_M varies with defect concentration and, therefore, with oxygen potential.

Fueki and Wagner[23] have overcome this problem by using the differential form of equation 3.80, with respect to the external oxygen partial pressure, since the value of p'_{O_2} is held constant by the equilibrium at the metal–scale interface, i.e.

$$\frac{dk'}{d \ln p''_{O_2}} = \frac{Z_c}{4} D_M \tag{3.81}$$

If, therefore, the value of k' is measured as a function of the external oxygen partial pressure, p''_{O_2}, at constant temperature, then D_M can be obtained from the slope of the plot of k' versus ln p''_{O_2}. Fueki and Wagner[23] determined the diffusion coefficient of nickel in NiO as a function of oxygen partial pressure in this way. The results indicated that D_{Ni} was proportional to $p''^{1/6}_{O_2}$ at 1000 °C, to $p''^{1/4}_{O_2}$ at 1300 °C, and to $p''^{1/3.5}_{O_2}$ at 1400 °C, indicating a predominance of doubly charged vacancies at 1000 °C and singly charged vacancies at 1300 °C with the existence of neutral vacancies at higher temperatures.

Mrowec and Przybylski have applied this method to the cobalt–oxygen system, since the defect structure is well known and the self diffusion coefficient of cobalt has been measured accurately using tracer techniques by Carter and Richardson[24] and also Chen *et al.*[25].

Discs of cobalt were oxidised between 900 and 1300 °C and between 10^{-6} and 1 atm of oxygen[21]. The resulting plot of k'' against log p_{O_2} was curved showing, according to equation 3.81, that k'' varied with oxygen partial pressure. The slope of the curve, $dk''/d \log p_{O_2}$, was used to obtain values of D_{Co} which give a convincing demonstration of the soundness of Wagner's analysis of parabolic oxidation as shown below.

Tracer results:

$D_{Co} = 0.0052 \exp(-19200/T)$ cm^2 s^{-1} (reference 24)
$D_{Co} = 0.0050 \exp(-19400/T)$ cm^2 s^{-1} (reference 25)

Oxidation measurements:

$D_{Co} = 0.0050 \exp(-19100/T)$ cm^2 s^{-1} (reference 22)

Thus insight into the transport mechanisms by which oxidation reactions proceed is obtained by studying the variation of the oxidation rates with

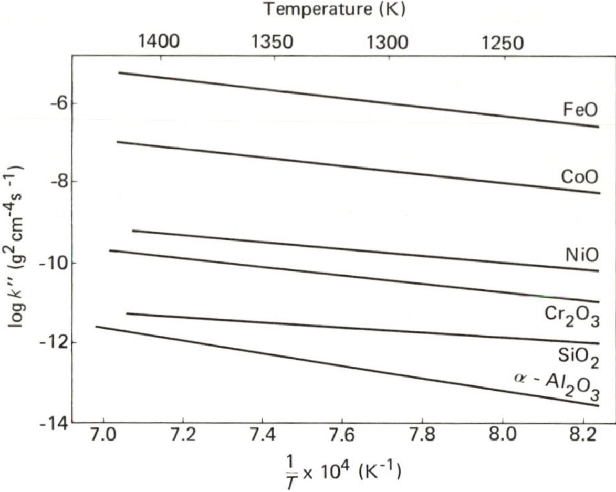

Fig. 3.13 Order-of-magnitude parabolic rate constants for the growth of several oxides

oxygen pressure and temperature. The absolute rates themselves are useful in comparing the oxidation kinetics of several metals. Some of these results are plotted in Fig. 3.13 from which it can be seen that the highest oxidation rates are, by and large, obtained in systems whose oxides have the highest defect concentrations. Clearly, in the development of oxidation resistant alloys, the aim is to evolve a composition that develops one of the slow growing oxides as a scale. This is usually done to encourage formation of Al_2O_3, SiO_2, or Cr_2O_3 but rarely, if ever, by taking advantage of doping, which would be unreliable.

Linear rate law

Under certain conditions the oxidation of a metal proceeds at a constant rate and is said to obey the 'linear rate law', i.e.

$$x = k_l t \tag{3.82}$$

where x is the scale thickness and k_l is the linear rate constant. The units of k_l depend upon the method used to follow the reaction. Where oxide scale thickness is measured, as in equation 3.82, then k_l has the units cm s^{-1}, if mass gain measurement were to be used then the appropriate linear rate constant would have units of g cm^{-2} s^{-1} and so on.

The linear rate law is usually observed under conditions where a phase boundary process is the rate determining step for the reaction, although other steps in the mechanism of oxidation can lead to the same result.

Principally, diffusion through the scale is unlikely to be rate limiting when the scale is thin, i.e. in the initial stages of the process. In this case, the metal–oxide and oxide–gas interfaces cannot be assumed to be in thermodynamic equilibrium, although there are no known cases where the reaction at the metal–scale interface has been found to play a rate determining role. It can, therefore, be assumed that the reactions occurring at the metal–scale interface, i.e. the ionisation of metal in the case of cation conductors or the ionisation of metal and oxide formation in the case of anion conducting scales, are fast and, in general, attention can be turned to the processes occurring at the scale–gas interface for an interpretation of the linear rate law.

At the scale–gas interface, the processes can be broken down into several steps. Reactant gas molecules must approach the scale surface and become adsorbed there, the adsorbed molecules then split to form adsorbed oxygen which eventually attracts electrons from the oxide lattice to become initially chemisorbed and finally incorporated into the lattice. The removal of electrons from the scale causes change in the concentration of electronic defects in the oxide at the scale–gas interface. These processes can be represented as follows.

$$\tfrac{1}{2}O_2\,(g) \underset{(1)}{\rightarrow} \tfrac{1}{2}O_2\,(ad) \underset{(2)}{\rightarrow} O\,(ad) \underset{(3)}{\overset{e^-}{\rightarrow}} O^-\,(chem) \underset{(4)}{\overset{e^-}{\rightarrow}} O^{2-}\,(latt) \tag{3.83}$$

Equation 3.83 is written in terms of the oxygen molecule being the active oxidising species in the gas phase. Except for cases of very low oxygen partial pressure, reactions carried out in such atmospheres do not lead to constant rate kinetics. It can, therefore, be assumed that the steps 2, 3, and 4, which occur on the oxide surface, are fast reactions.

If, however, the oxidising medium is a CO–CO_2 mixture then constant rate kinetics are readily observed. In this case the surface reactions may be represented as follows.

$$\begin{array}{ccccccc} & & & & CO\,(g) & & \\ & & & & \downarrow (5) & & \\ CO_2\,(g) & \underset{(1)}{\rightarrow} CO_2\,(ad) & \underset{(2)}{\rightarrow} & CO\,(ad) & +\; O\,(ad) & \underset{(3)}{\overset{e^-}{\rightarrow}} O^-\,(chem) & \underset{(4)}{\overset{e^-}{\rightarrow}} O^{2-}\,(latt) \\ & & & \downarrow & \downarrow & & \\ & & & CO\,(g) & CO_2\,(g) & & \end{array} \tag{3.84}$$

Primarily, CO_2 adsorbs on to the oxide surface and there dissociates to adsorbed CO and O species. The adsorbed oxygen then goes through the ionisation stages already given in equation 3.83. Two further processes are the desorption of adsorbed carbon monoxide and the possibility of the removal of adsorbed oxygen by reaction with carbon monoxide molecules from the gas phase. Since steps 3 and 4 can be assumed from above to be fast and since the rate of impingement of CO_2 on the scale surface, from atmospheres containing high concentrations of CO_2, is high, the most likely rate determining step is step 2 (dissociation of adsorbed carbon dioxide to form adsorbed carbon monoxide molecules and oxygen atoms on the surface of the oxide scale).

Hauffe and Pfeiffer[26] investigated the oxidation of iron in CO–CO_2 mixtures between 900 and 1000 °C and found that, under a total gas pressure of 1 atmosphere, the rate was constant and proportional to $(p_{CO_2}/p_{CO})^{2/3}$, i.e. to $p_{O_2}^{1/3}$. They suggested that this may indicate that chemisorption of oxygen is the rate controlling step.

The same reaction was investigated in more detail by Pettit *et al.*[27] who assumed that the rate determining step was the dissociation of CO_2 to give CO and adsorbed oxygen according to

$$CO_2\,(g) = CO\,(g) + O\,(ad) \tag{3.85}$$

If this assumption is correct, then the rate of the reaction per unit area, $\dot{n}/A$, would be given by

$$\frac{\dot{n}}{A} = k'_{p_{CO_2}} - k''_{p_{CO}} \tag{3.86}$$

when k' and k'' are constants specifying the rates of the forward and reverse reactions respectively. During the initial stages of reaction, at least, the scale is largely in equilibrium with the iron substrate, thus substantial changes in oxide defect concentrations at the scale–gas interface do not occur. Consequently, this factor plays no part in equation 3.86.

If the gas had a composition in equilibrium with iron and wustite, then the reaction rate would be zero, i.e. $\dot{n}/A = 0$.

Consequently

$$\frac{k''}{k'} = \frac{p_{CO_2}}{p_{CO}} K \tag{3.87}$$

and, therefore, from equations 3.86 and 3.87, the rate under CO_2/CO ratios exceeding the equilibrium condition is given by

$$\frac{\dot{n}}{A} = k'\,(p_{CO_2} - Kp_{CO}) \tag{3.88}$$

The component partial pressures can be written in terms of the relevant mol fractions N_{CO_2} and N_{CO} and the total pressure P.

$$p_{CO_2} = PN_{CO_2} \tag{3.89}$$

$$p_{CO} = PN_{CO} = P(1 - N_{CO_2}) \tag{3.90}$$

Substituting equations 3.89 and 3.90 in equation 3.88 gives

$$\frac{\dot{n}}{A} = k'P[(1 + K)N_{CO_2} - K]$$

which can be rearranged to

$$\frac{\dot{n}}{A} = k'P(1 + K)[N_{CO_2} - N_{CO_2}(\text{eq})] \quad (3.91)$$

where N_{CO_2} (eq) refers to the mol fraction of CO_2 in the CO_2/CO mixture in equilibrium with iron and wustite. According to equation 3.91 the oxidation rate at a given temperature should be proportional to both the mol fraction of CO_2 in the gas and the total pressure of the gas.

The experimental results for 925 °C shown in Fig. 3.14 substantiate that both of the above conditions are met and, therefore, the rate determining step in the reaction is the decomposition of carbon dioxide to carbon monoxide and absorbed oxygen.

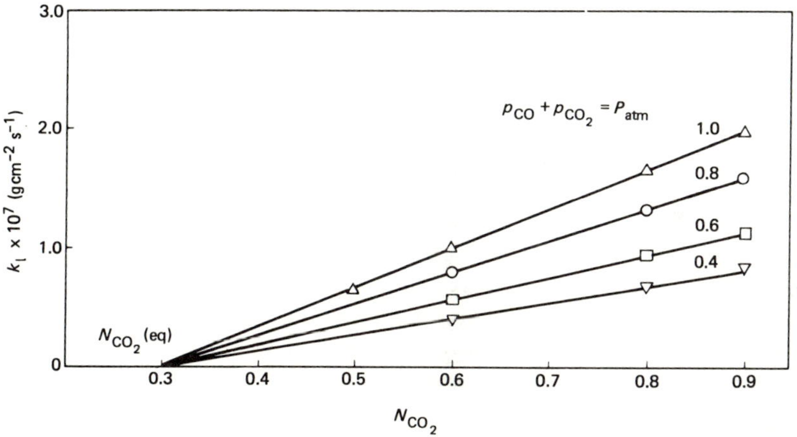

Fig. 3.14 Variation of oxidation rate of iron in CO–CO_2 atmospheres with CO_2 mol fraction, *N*. After Pettit *et al.*[27]

Rarefied atmospheres

Constant reaction rates can also be observed during reactions of a metal under rarefied atmospheres and with reactive gases diluted with an inert, unreactive gas.

A good example of this case is the heat treatment of metals under partial vacuum when the atmosphere consists of nitrogen and oxygen molecules at a low pressure, where the rate determining step is likely to be the rate of impingement of oxygen molecules on the specimen surface. This rate can be calculated by using the Hertz–Knudsen–Langmuir equation derived from the kinetic theory of gases

$$j_{O_2} = \alpha \frac{p_{O_2}}{(2\pi M_{O_2} RT)^{1/2}} \tag{3.92}$$

where p_{O_2} is the oxygen partial pressure in the partial vacuum
M_{O_2} is the relative molecular mass of oxygen
R is the gas constant
T is absolute temperature
α is the adherence coefficient

α has a maximum value of unity when all of the impinging molecules adsorb and enter into reaction. Thus the maximum possible reaction rate is given when $\alpha = 1$ in equation 3.92. When using equation 3.92 it is necessary to use compatible units and these are tabulated below.

Flux j_{O_2}	Pressure p_{O_2}	R	M_{O_2}
mol cm^{-2} s^{-1}	atm	cc atm mol^{-1} deg^{-1}	g mol^{-1}
mol cm^{-2} s^{-1}	dyn cm^{-2}	erg mol^{-1} deg^{-1}	g mol^{-1}
mol m^{-2} s^{-1}	N m^{-2}	N m mol^{-1} deg^{-1}	kg mol^{-1}

Evaluating equation 3.92 for reaction of a metal with oxygen existing at a partial pressure of 10^{-6} atm in a partial vacuum at 1000 K, the maximum possible reaction rate is $j_{O_2} = 2.45 \times 10^{-7}$ mol cm^{-2} s^{-1} or 7.8×10^{-6} g cm^{-2} s^{-1}.

According to equation 3.92, the reaction rate will vary with temperature, for a given oxygen partial pressure, according to

$$j_{O_2} = \text{const. } T^{-1/2} \tag{3.93}$$

Dilute gases

When a metal is reacted with an atmosphere consisting of an active species diluted by an inactive or inert species, the active molecules are rapidly depleted in the gas layers immediately adjacent to the specimen surface. Subsequent reaction can then only proceed as long as molecules of the active species can diffuse through the denuded layer to the metal surface. When the concentration of the active species is low in the atmosphere this step can become rate controlling. The denuded gas layer next to the metal surface may be regarded as a boundary layer of thickness δ. If the concentrations of the active species at the metal surface and in the bulk gas are expressed, respectively, by the partial pressures p_i'' and p_i' then the rate at which active species, i, diffuse across the boundary layer may be expressed as

$$j_i = -D_i \frac{p_i' - p_i''}{\delta} \tag{3.94}$$

which represents the maximum possible reaction rate observable under these circumstances. In most cases, at least at the beginning of the reaction, the value of p_i'' is very low and can be neglected giving

$$j_i \approx -D_i \frac{p_i'}{\delta} \tag{3.95}$$

i.e. the reaction rate is directly proportional to p_i', the partial pressure of the active species in the bulk atmosphere. The sensitivity of the reaction rate to temperature arises solely from the way in which D_i and, to a lesser extent, δ, vary with temperature.

Reaction rate control by this step can be identified most securely by the sensitivity of the reaction rate to the gas flow rate. If the gas flow rate is increased the thickness of the boundary layer, δ, is decreased and, correspondingly, the reaction rate increases according to equation 3.94.

It is important to note that the concentration of adsorbed species and, therefore, the rates of surface reactions involving adsorbed species also vary with the partial pressure of the active species in the gas phase. Thus, in order to identify boundary layer transport as the process controlling a constant rate, sensitivity of the rate to *both* active species partial pressure *and* gas flow rate must be established.

Transition from linear law to parabolic law

It was stated earlier that constant rate kinetics are observed with thin scales. This is obviously a relative term and denotes the initial stages of a reaction which may continue for a long, or short, time depending on the conditions under which the reaction occurs.

Initially, when a scale is formed and is very thin, diffusion through the scale will be rapid establishing virtual equilibrium with the metal at the scale–gas interface. In other words, the metal activity at this interface will be maintained at a high value, initially close to unity, by rapid diffusion within the scale. Under these conditions, the rate of the reaction is likely to be controlled by one of the steps discussed above. As the reaction proceeds at constant rate the scale layer thickens, at the same time the flux of ions through the scale must be equivalent to the surface reaction rate. To maintain this constant flux, the activity of the metal at the scale–gas interface must fall as the scale thickens, eventually approaching the value in equilibrium with the atmosphere. Since the metal activity cannot fall below this value, further increase in scale thickness must result in a reduction in the metal activity gradient across the scale and, consequently, to a reduction in ionic flux and the reaction rate. At this point the transport of ions across the scale becomes the rate controlling process and the rate falls with time according to a parabolic rate law.

Logarithmic rate laws

When metals are oxidised under certain conditions, typically at low temperatures of up to about 400 °C, the initial oxide formation, up to the 1000 Å range, is characterised by an initial rapid reaction that quickly reduces to a very low rate of reaction. Such behaviour has been found to conform to rate laws described by logarithmic functions such as:

$$x = k_{\log} \log(t + t_0) + A \quad \text{(direct log law)} \tag{3.96}$$

and

$$1/x = B - k_{\text{il}} \log t \quad \text{(inverse log law)} \tag{3.97}$$

where A, B, t_0, $k_{\log}$, and k_{il} are constants at constant temperature.

Several interpretations of this type of behaviour have been provided, for a detailed description and summary, Kofstad[28] and Hauffe[29] should be consulted. The phenomenon being principally one that occurs at low temperatures, a strictly qualitative description of the various interpretations will be given here, since this phase of the oxidation reaction generally occurs during the heating period of an investigation into the formation of thick scales at high temperatures.

Interpretations of the logarithmic laws have been based on the adsorption of reactive species, the effects of electric fields developed across oxide layers, quantum-mechanical tunnelling of electrons through the thin scales, progressive blocking of low resistance diffusion paths, non-isothermal conditions in the oxide layer, and nucleation and growth processes. A concise summary of these theories has been given by Kofstad[28].

Two things should be borne in mind. Firstly, it is difficult experimentally to obtain accurate and reliable measurements of reaction rates relating to the formation of the initial oxide layers, although, lately, modern high vacuum techniques may allow an accurate start to the reaction to be assumed by achieving thermal equilibrium before admitting the reactant gas. Secondly, in any thermodynamic treatment carried out to assess defect concentrations or concentration gradients, it may be necessary to include surface, or interfacial, energies when considering such thin films, thus the actual deviations from stoichiometry in the film may be unknown, similarily the physical properties of the film may vary significantly from those of the bulk material. Due to difficulties such as these, the interpretation of logarithmic rate behaviour is the least understood area of metal oxidation.

Adsorption has been assumed to be the rate determining process during early oxide film formation. When a clean surface is exposed to an oxidising gas, each molecule impinging on the surface may either rebound or adsorb. The fraction, α, that remains adsorbed on the metal surface should be constant for a constant temperature and oxygen partial pressure. Therefore, under these conditions a constant reaction rate is expected. However, the value of α is markedly lower on those parts of the surface covered with a monolayer or oxide nuclei, thus, as adsorption or oxide nucleation proceeds, the reaction rate is expected to decrease accordingly until complete coverage of the surface by oxide has been achieved when a much lower rate is expected to be observed.

These initial processes of adsorption and nucleation have been investigated principally by Bénard and co-workers[30], who have demonstrated convincingly that the initial nucleation of oxide occurs at discrete sites on the metal surface. The oxide islands proceed to grow rapidly over the metal surface until complete coverage is eventually achieved. Furthermore, it has been demonstrated that the step of adsorption begins at oxygen partial pressures significantly below the decomposition pressure of the oxide.

When the oxidising gas chemisorbs on to the oxide surface growing on a metal, it captures, or withdraws, electrons from the oxide lattice according to

$$\tfrac{1}{2}O_2(g) \rightarrow O(ad) \xrightarrow{e^-} O^-(chem) \tag{3.83}$$

The result is to create additional electron holes in the surface of a p-type oxide or reduce the number of excess electrons in the surface of an n-type oxide. The electrons are withdrawn from a depth of about 100 Å over which, consequently, a strong electric field exists known as a space charge. This concept has been used by Cabrera and Mott[31] to explain the inverse logarithmic law for the case of the formation of thin oxide films of very tight stoichiometry and, consequently, low ionic migration rates at low temperatures due to the low defect concentrations (the electronic mobility is assumed to be greater than that of the ions). The electric field set up by chemisorption of oxygen, aided by quantum-mechanical tunnelling of electrons from metal through the oxide to the adsorbed oxygen, acts on the mobile ionic species and increases the migration rate. Clearly, since the potential difference between the metal and adsorbed oxygen species is assumed to be constant, the thinner the oxide layer the stronger the field and the faster the ionic diffusion. As the film thickens, the field strength is reduced and the reaction rate falls. When the oxide layer thickness exceeds about 100 Å, tunnelling of electrons from the metal through the scale is not feasible and the full potential difference is no longer exerted. Thus, eventually, very low reaction rates are observed.

Inverse logarithmic rates have been reported by Young *et al.*[32] for copper at

70 °C, Hart[33] for aluminium at 20 °C and Roberts[34] for iron at temperatures up to 120 °C.

Mott[35] proposed an interpretation of the direct log law, later revised by Hauffe and Ilschner[36], by explaining that the initial rapid reaction rates were due to quantum-mechanical tunnelling of electrons through the thin oxide layer. Although electrons are not generally mobile in oxides at low temperatures, the electrical conductivity of thin oxide layers on metals is, surprisingly, much higher than would be expected from bulk electrical properties. Thus, if the transport of electrons were the rate determining step, the reaction rate would be surprisingly high in the initial stages when the film is very thin. As the film thickens, the maximum thickness for which tunnelling of electrons is valid is exceeded and the observed reaction rate falls away markedly.

Other interpretations of log laws have rejected the assumption of uniform transport properties across the oxide layers and assume instead that paths for rapid ionic transport exist, e.g. along grain boundaries, or dislocation pipes, or over the surface of pores. As growth proceeds these paths are progressively blocked, either by recrystallisation and grain growth, or by blocking of pores by growth stresses in the oxide, or by a combination of all three processes. It is also possible that pores may act as vacancy sinks in the oxide and grow into cavities large enough to interfere effectively with ionic transport by reducing the cross-section of compact oxide through which transport may occur. The formation of second phase alloy oxides of low ionic conductivity may also give the same progressive resistance to ionic migration. These theories have been considered principally by Evans and co-workers[37].

In view of the multiple interpretations available, it is clearly wise in practice to investigate thoroughly and judge each case on its merits rather than pass verdict from a simple examination of kinetics alone.

The presentation of Wagner's treatment of high temperature oxidation of metals given here follows the original very closely. One of the assumptions made in this treatment is that the oxide scale under consideration is thick compared with the distance over which space charge effects, discussed above, occur. Fromhold[38,39,40] has recently considered the effect of the presence of space charges at the scale–gas and metal–scale interfaces. His conclusions show that their presence would cause the rate constant to vary with scale thickness in the range 5000–10 000 Å. This analysis is not considered further here since the oxide scales normally encountered are very much thicker than this.

Relationships between parabolic rate constants

In the above treatment the parabolic rate constant is derived for the units relating to measurement of the oxide thickness as the reaction parameter. As mentioned before, there are several methods of following the reaction, depending on choice of reaction parameter, each of which produces its own particular parabolic rate constant, as shown below using Wagner's[14] notation for the various parabolic rate constants.

(*a*) *Measurement of scale thickness* (x)

$$\frac{dx}{dt} = \frac{k'}{x} \quad \text{i.e. } x^2 = 2k't \tag{3.98}$$

k' is the 'practical tarnishing constant' or 'scaling constant' and has units of $cm^2\ s^{-1}$.

(*b*) *Measurement of the mass increase of the specimen* (m)
The parabolic rate constant k'' is defined by

$$\left(\frac{m}{A}\right)^2 = k''t \tag{3.98}$$

where A is the area over which reaction occurs.
k'' is also referred to as the 'practical tarnishing' or 'scaling constant' and has units of $g^2\ cm^{-4}\ s^{-1}$.

(*c*) *Measurement of metal surface displacement* (l)
Measuring the thickness of metal consumed leads to the relationship defining k_c

$$l^2 = 2k_ct \tag{3.99}$$

k_c is called the 'corrosion constant' and has units of $cm^2\ s^{-1}$.

(*d*) *Rate of growth of scale of unit thickness*
The relevant rate constant is defined as the rate of growth over unit area in equivalents per second of a scale of unit thickness, i.e.

$$k = \frac{x}{A}\frac{d\bar{n}}{dt} \tag{3.100}$$

where $\bar{n}$ is the number of equivalents in the oxide layer of thickness x.
k is called the 'theoretical tarnishing constant' and has units of equivalents $cm^{-1}\ s^{-1}$. This is sometimes called the rational rate constant and denoted k_r.

It is easy to see that having so many different ways of expressing parabolic rate constants, and so many different symbols, produces a situation ripe for confusion. Consequently, it is necessary to check the definition of a rate constant very carefully when evaluating quantitative data.

It is easy to calculate the value of any rate constant from any other since they all represent the same process. The relationships are given in Table 3.1.

Table 3.1 Relationships between variously defined parabolic oxidation rate constants

A / B	k' cm^2 s^{-1}	k'' g^2 cm^{-4} s^{-1}	k_c cm^2 s^{-1}	k eq cm^{-1} s^{-1}
k' cm^2 s^{-1}	1	$2[M_X/\bar{V}Z_X]^2$	$[\bar{V}_M/\bar{V}]^2$	$1/\bar{V}$
k'' g^2 cm^{-4} s^{-1}	$\frac{1}{2}[\bar{V}Z_X]^2/M_X^2$	1	$\frac{1}{2}[\bar{V}Z_X/M_X]^2$	$\bar{V}/2[Z_X/M_X]^2$
k_c cm^2 s^{-1}	$[\bar{V}/\bar{V}_M]^2$	$2[M_X/\bar{V}_M Z_X]^2$	1	$\bar{V}/\bar{V}_M^2$
k eq cm^{-1} s^{-1}	$\bar{V}$	$2/\bar{V}[M_X/Z_X]^2$	$\bar{V}_M^2/\bar{V}$	1

Notes:
The relating factor F is given according to

$$A = FB$$

A is listed horizontally and B vertically.

The symbols have the following meaning:

$\bar{V}$ = equivalent volume of scale
$\bar{V}_M$ = equivalent volume of metal
M_X = relative atomic mass of non-metal X (oxygen, sulphur, etc.)
Z_X = valency of X

References

1 Kröger, F. O., *The Chemistry of Imperfect Crystals*, North Holland Publishing Co., Amsterdam, 1964
2 Kofstad, P., *Nonstoichiometry, Diffusion, and Electrical Conductivity in Binary Metal Oxides*, Wiley, New York, 1972
3 von Baumbach, H. H. and Wagner, C., *Z. phys. Chem.*, **22,** 199, 1933
4 Mohanty, G. P. and Azaroff, L. V., *J. Chem. Phys.*, **35,** 1268, 1961
5 Thomas, D. G., *J. Phys. Chem. Solids*, **3,** 229, 1957
6 Scharowsky, E., *Z. Physik*, **135,** 318, 1953
7 Hoffmann, J. W. and Lauder, I., *Trans. Farad. Soc.*, **66,** 2346, 1970
8 Bransky, I. and Tallan, N. M., *J. Chem. Phys.*, **49,** 1243, 1968
9 Eror, N. G. and Wagner, J. B., *J. Phys. Stat. Sol.*, **35,** 641, 1969
10 Mitoff, S. P., *J. Phys. Chem.*, **35,** 882, 1961
11 Meier, G. H. and Rapp, R. A., *Z. phys. Chem. N.F.*, **74,** 168, 1971
12 Koel, G. J. and Gellings, P. J., *Oxid. Metals.*, **5,** 185, 1972
13 Pope, M. C. and Birks, N., *Corros. Sci.*, **17,** 747, 1977
14 Wagner, C., *Z. phys. Chem.*, **21,** 25, 1933

15 Fisher, B. and Tannhauser, D. S., *J. Chem. Phys.*, **44,** 1663, 1966
16 Eror, N. G. and Wagner, J. B., *J. Phys. Chem. Solids*, **29,** 1597, 1958
17 Carter, R. E. and Richardson, F. D., *TAIME*, **203,** 336, 1955
18 Bridges, D. W., Baur, J. P. and Fassell, W. M., *J. Electrochem. Soc.*, **103,** 619, 1956
19 Mrowec, S. and Stoklosa, A., *Oxid. Metals*, **3,** 291, 1971
20 Mrowec, S., Stoklosa, A. and Godlewski, K., *Crystal Lattice Defects*, **5,** 239, 1974
21 Mrowec, S. and Przybylski, K., *Oxid. Metals*, **11,** 365, 1977
22 Mrowec, S. and Przybylski, K., *Oxid Metals*, **11,** 383, 1977
23 Fueki, K. and Wagner, J. B., *J. Electrochem Soc.*, **112,** 384, 1965
24 Carter, R. E. and Richardson, F. D., *J. Metals*, **6,** 1244, 1954
25 Chen, W. K., Peterson, N. L. and Reeves, W. T., *Phys. Rev.*, **186,** 887, 1969
26 Hauffe, K. and Pfeiffer, H., *Z. Elektrochem.*, **56,** 390, 1952
27 Pettit, F. S., Yinger, R. and Wagner, J. B., *Acta Met.*, **8,** 617, 1960
28 Kofstad, P., *High Temperature Oxidation of Metals*, Wiley, New York, 1966
29 Hauffe, K., *Oxidation of Metals*, Plenum Press, New York, 1965
30 Bénard, J., *Oxydation des Méteaux*, Gautier–Villars, Paris, 1962
31 Cabrera, N. and Mott, N. F., *Rept. Progr. Phys.*, **12,** 163, 1948
32 Young, F. W., Cathcart, J. V. and Gwathmey, A. T., *Acta Met.*, **4,** 145, 1956
33 Hart, R. K., *Proc. Roy. Soc.*, **236A,** 68, 1956
34 Roberts, M. W., *Trans. Farad. Soc.*, **57,** 99, 1961
35 Mott, N. F., *Trans. Farad. Soc.*, **36,** 472, 1940
36 Hauffe, K. and Ilschner, B., *Z. Elektrochem.*, **58,** 382, 1954
37 Evans, U. R., *The Corrosion and Oxidation of Metals*, Edward Arnold, London, 1960
38 Fromhold, A. T., *J. Phys. Chem. Solids*, **33,** 95, 1972
39 Fromhold, A. T. and Cook, E. L., *Phys. Rev.*, **175,** 877, 1968
40 Fromhold, A. T., *J. Phys. Soc. Japan*, **48,** 2022, 1980

4
Oxidation of pure metals

Introduction

The following discussion will necessarily be idealised since it is difficult to find systems which conform completely to any single mode of rate control. This makes it correspondingly difficult to choose examples to illustrate completely any single rate controlling process. Even such factors as specimen shape can influence the detailed mechanism operating over the life of the specimen. For example, when a metal is exposed to an oxidising atmosphere at high temperature the initial reaction is expected to be very rapid since the oxide layer formed is very thin. However, if the parabolic rate law is extrapolated to zero scale thicknesses, an infinite rate is predicted. Clearly this is not so and the initial stage of oxidation must be controlled by some process other than ionic transport through thin oxide scales, e.g. the processing of gas molecules on the oxide surface according to

$$O_2(g) = O_2(ad) = 2O(ad) \rightleftharpoons 2O^-(chem) + h \rightarrow 2O^{2-}(latt) + 2h$$

This describes the adsorption, dissociation, chemisorption, and ionisation of oxygen and would lead to a constant reaction rate if rate controlling. However, these processes are so rapid that the oxidation period over which they control the reaction rate is rarely observed. In fact, observation of this early period is extremely difficult since, under most conditions, the oxide scale formed during the period of heating the sample to temperature is usually sufficiently thick for ionic diffusion through the scale to be rate controlling by the time that reaction monitoring is begun under isothermal conditions.

The initial, surface processing controlled stage can, however, be observed under special conditions. This has been illustrated by Pettit, Yinger and Wagner (described previously) who chose a gas system with a very slow surface processing reaction. Alternatively, it may be possible to heat the specimen in an inactive atmosphere or vacuum and suddenly introduce the oxidising species when isothermal conditions have been reached. No results of this type of experiment have been reported.

Once the oxidation of the metal reaches the stage where ionic diffusion is rate controlling, a parabolic rate law is found to hold for a period whose duration depends upon factors such as specimen geometry and scale mechanical properties.

Specimen geometry is important because, as the reaction proceeds, the metal core gets smaller and so the area of metal–scale interface gets smaller. Thus, when the reaction rate is expressed in mass gain per unit area, taking the initial surface area of the metal specimen to be constant can lead to small but significant errors. This aspect has been thoroughly treated by Romanski[1], Bruckmann[2], and Mrowec and Stoklosa[3] who showed that, unless corrections allowing for the reduction of the scale–metal interface area are made, the rate of the reaction will apparently fall below that predicted on the basis of the parabolic rate law.

The above consideration applies so long as the scale and metal maintain contact at the scale–metal interface. As the oxidation reaction proceeds, however, stresses are generated within the oxide, and these are called growth stresses. These stresses will be discussed in detail in the next chapter but a brief discussion is needed at this point. In systems where cations are mobile, growth stresses arise in the scale since the scale must relax to maintain contact with the metal as the metal atoms cross the scale–metal interface to diffuse outwards as cations and electrons. If the scale does not relax, voids will form at the scale–metal interface separating the scale from the metal. On a planar interface there are no forces restraining such relaxation, but at edges and corners it is not possible for the scale to relax in both, or all three, directions. The geometry of the scale in these regions is stabilised and resists such relaxation in much the same way as the sides of a cardboard box cannot be flexed in the corner regions. In these geometrically stabilised regions, the scale must creep to maintain contact with the metal at a rate that is determined by the oxidation rate of the metal, i.e. the rate at which metal is being removed. The adhesion between scale and metal is the maximum force that can be exerted to cause the scale to creep and maintain contact. Thus, unless the scaling rate is low or the scale is quite plastic, contact between scale and metal will be lost as the reaction proceeds. As the scale thickens, contact will eventually be lost in any real situation involving corners. With cylindrical specimens a similar situation arises.

As oxidation proceeds and contact between scale and metal is lost at edges and corners, spreading over the faces with time, the area across which cations can be supplied is therefore decreased. This means that, to supply the scale–gas interface over the separated areas, the cations have a longer diffusion distance (see Fig. 4.1) and hence the reaction rate falls. This leads to specific reaction rates that are lower than expected when the original metal surface area is used in the calculation.

Frequently this loss of contact causes the formation of a porous zone of scale between the outer compact layer and the metal. The mechanism of formation is quite straightforward as outlined below and shown in Fig. 4.2. When scale separation occurs, the metal activity at the inner surface of the scale is high and so cations still continue to migrate outwards. However, this causes the metal activity to fall and the oxygen activity to rise

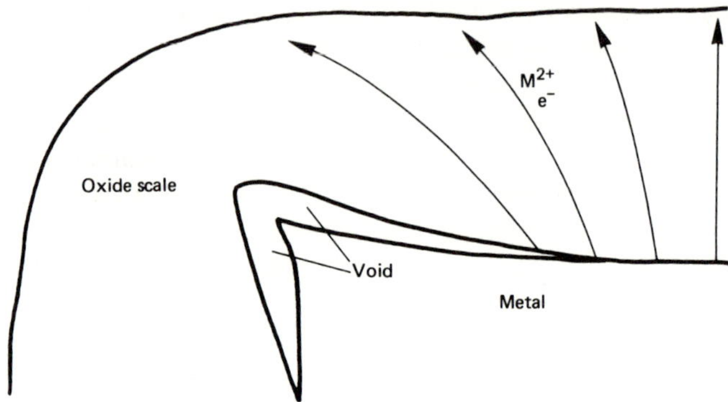

Fig. 4.1 Schematic diagram showing longer diffusion distance for cations to supply scale–gas interface when scale–metal separation occurs at edges and corners

correspondingly. As the oxygen activity rises, the oxygen partial pressure in local equilibrium with the scale inner surface rises and oxygen 'evaporates' into the pore, diffuses across the pore, and forms oxide on the metal surface. In this way a porous layer can be formed next to the scale–metal interface while maintaining outward cation diffusion. The oxide scale dissociates more rapidly at grain boundaries than at grain surfaces.

The above discussion shows that physical aspects of the scale and scale–metal interface configuration can markedly influence the apparent kinetic behaviour and rate appearance. It cannot be emphasised too strongly that careful metallographic and other examination is essential for accurate and correct interpretation of observed kinetics. These factors also indicate that specimen size may influence the course of the kinetics. Thus, extrapolating from results on small specimens to predict results on large pieces or pieces of

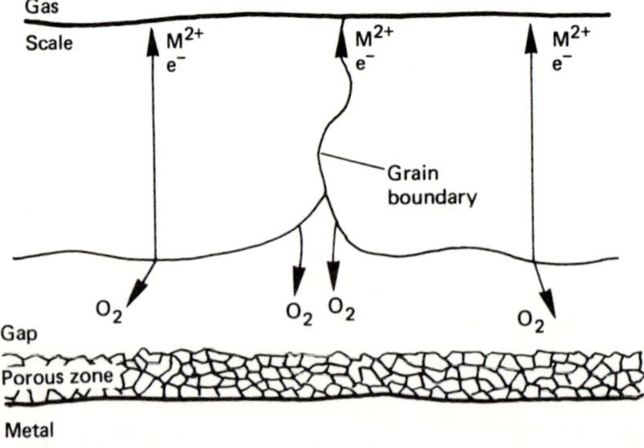

Fig. 4.2 Mechanism of formation of porous zone due to scale–metal separation. Note that the scale decomposes more readily at the grain boundaries in the oxide scale than at the grain surfaces

different geometry may not be easy or feasible. The above and other aspects will be brought out in the following discussion of the oxidation of selected pure metals.

Systems forming single layer scales

Oxidation of nickel

Nickel is an ideal metal for oxidation studies since, under normal temperature and pressure conditions, it forms only one oxide, NiO, which is a p-type semiconductor with cation deficit. The mechanism by which oxidation of nickel proceeds is, therefore, expected to involve simply the outward migration of cations and electrons forming a single phase scale.

Early measurements of the oxidation of nickel over the temperature range 700–1300 °C showed a surprising variation of the parabolic rate constant, which varied over four orders of magnitude[4–6]. Later determinations using nickel of significantly improved purity[7,8] showed that the parabolic rate constant for nickel containing ~0.002% impurity was reliably reproducible and was lower than that for the less pure nickel samples used earlier. Since the majority of impurity elements found in nickel are either divalent or trivalent, they will, when dissolved in the oxide, either have no effect on, or increase, the mobility of cations in NiO. It is, therefore, understandable that the impure nickel specimens were found in general to oxidise more rapidly than pure nickel. The activation energy of the process remained fairly constant regardless of impurity level indicating that the rate determining step remained unchanged. The scale morphology has also been found to vary significantly with the purity of the metal. This is particularly well demonstrated by results obtained at 1000 °C[9].

Using nickel of high purity a compact, adherent, single layer scale of NiO is formed. Platinum markers placed on the metal surface before the reaction are found afterwards at the scale–metal interface. This is very strong evidence that the scale forms entirely by the outward migration of cations and electrons.

Earlier work using nickel which contained low concentrations of impurities (0.1%) showed that a fundamentally different scale was produced. In this case the scale consisted of two layers, both of NiO, the outer compact and the inner porous. Furthermore, platinum markers were found to lie at the interface between 'the outer compact and inner porous NiO layer after oxidation'[10–12].

The presence of the markers at the boundary between the compact and porous layers of the scale implies that the outer layer grows by outward cation migration and that the inner porous layer grows by inward migration of oxygen. The mechanism proposed to explain the formation of such a scale[10] assumes that the oxide is separated from the metal early in the oxidation process due to insufficient oxide plasticity, a situation which is aggravated by higher scaling rates of the impure nickel. Once the scale loses contact with the metal, the oxygen activity at the inner surface rises and the scale dissociates at a corresponding rate. The oxygen migrates across the pore or gap and begins to form new nickel oxide on the metal surface. Birks and Rickert[10] showed that the oxygen partial pressures likely to be involved in this mechanism were sufficient to account for the observed reaction rate. A diagrammatic representation of this process is the same as that given in Fig. 4.2.

In addition, Mrowec and Werber[13] propose that, due to oxide dissociation being faster at grain boundaries than over grain faces, the outer compact layer is penetrated by microcracks in the grain boundaries and that, consequently, inward transport of oxygen molecules from the atmosphere also plays a role.

As discussed earlier, NiO is a p-type cation deficient semiconductor and, therefore, the cations will migrate with electrons from the scale–metal interface to the scale–gas interface during oxidation. Correspondingly, there will be a flow of defects, cation vacancies, and electron holes in the opposite direction. Consequently, the driving force for the reaction will be reflected by the concentration gradient of cation vacancies across the scale. The nickel vacancies are formed according to

$$\tfrac{1}{2}O_2 = O_O + 2h\cdot + V_{Ni}'' \tag{4.1}$$

from which

$$C_{V_{Ni}''} = \text{const.}\, p_{O_2}^{1/6} \tag{4.2}$$

If associated vacancies V_{Ni}' are important the reaction would be

$$\tfrac{1}{2}O_2 = O_O + h\cdot + V_{Ni}' \tag{4.3}$$

from which

$$C_{V_{Ni}'} = \text{const.}\, p_{O_2}^{1/4} \tag{4.4}$$

The concentration gradient of cation vacancies across the scale is very sensitive to the oxygen partial pressure of the atmosphere as shown diagrammatically in Fig. 4.3. Correspondingly, the oxidation rate constant is expected to increase as the oxygen partial pressure of the gas is increased.

Fueki and Wagner[14] oxidised nickel over a range of oxygen partial pressures from the dissociation pressure of NiO to 1 atm at temperatures between 900 and 1400 °C. Their results showed that the observed parabolic rate constant did in fact vary with the oxygen pressure of the atmosphere; the value of n varying from lower than 6 at 1000 °C to 3.5 at 1400 °C. They also determined the self diffusion coefficient of nickel ions in nickel oxide, from their results, as a function of oxygen partial pressure. The values of the self diffusion coefficients were independent of oxygen partial pressure in the

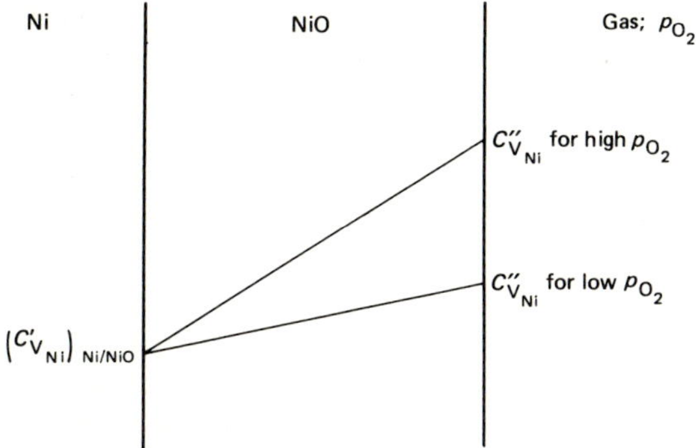

Fig. 4.3 Variation of cation vacancy concentration across a NiO scale for high and low oxygen partial pressures

low oxygen pressure range and proportional to $p_{O_2}^{1/6-1/3.5}$ in the intermediate oxygen pressure range.

They suggested that the values of n of 6 and 3.5 are consistent with the formation of doubly and singly charged vacancies according to equations 4.1 and 4.2 above. However, considering the results described later for the oxidation of cobalt, it is possible that intrinsic defects may also be involved.

Oxidation of zinc

Zinc also forms one oxide ZnO and, therefore, a single phase, single layered scale is expected when pure zinc is oxidised. However, ZnO is an n-type cation excess semiconductor, i.e. having interstitial Zn ions and electrons within the conduction band.

According to the equilibria established for this defect structure, the concentration gradient of interstitial zinc ions across the scale will depend also upon the oxygen partial pressure of the atmosphere, i.e. for divalent interstitials formed according to

$$ZnO = Zn_i^{\cdot\cdot} + 2e' + \tfrac{1}{2}O_2 \tag{4.5}$$

$$C_{Zn_i^{\cdot\cdot}} = \text{const.}\ p_{O_2}^{-1/6} \tag{4.6}$$

and for monovalent interstitials formed according to

$$ZnO = Zn_i^{\cdot} + e' + \tfrac{1}{2}O_2 \tag{4.7}$$

$$C_{Zn_i^{\cdot}} = \text{const.}\ p_{O_2}^{-1/4} \tag{4.8}$$

In both cases, increasing the oxygen partial pressure reduces the concentration of defects. Thus, the difference in concentration of interstitial zinc ions can be represented as

$$C^0_{\mathrm{Zn_i}} - C_{\mathrm{Zn_i}} = \text{const.}\,[(p^0_{\mathrm{O_2}})^{-1/n} - (p_{\mathrm{O_2}})^{-1/n}] \qquad (4.9)$$

$$= \text{const}'.\left[1 - \left(\frac{p_{\mathrm{O_2}}}{p^0_{\mathrm{O_2}}}\right)^{-1/n}\right]$$

where $C^0_{\mathrm{Zn_i}}$ represents the concentration at $p^0_{\mathrm{O_2}}$, the oxygen partial pressure in equilibrium with Zn and ZnO, i.e. at the scale–metal interface, $C_{\mathrm{Zn_i}}$ and $p_{\mathrm{O_2}}$ represent conditions at the scale–gas interface, and n is either 6 or 4. For practical situations $p_{\mathrm{O_2}} \gg p^0_{\mathrm{O_2}}$, the value $(p_{\mathrm{O_2}}/p^0_{\mathrm{O_2}})^{-1/n}$ is very small compared with unity, and therefore the gradient of concentration of interstitial zinc ions across the scale is insensitive to the oxygen partial pressure in the atmosphere. Consequently, the scaling rate constant for zinc will be insensitive to the external oxygen partial pressure so long as this is high compared with the value for the Zn/ZnO equilibrium. The situation is shown in Fig. 4.4.

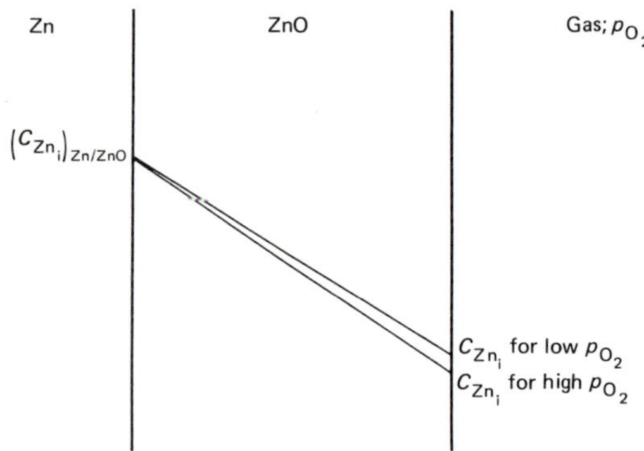

Fig. 4.4 Variation of concentration of singly or doubly charged interstitial zinc ions across a ZnO scale for high and low oxygen partial pressures

The independence of the parabolic rate constant for zinc in oxygen at 390 °C was demonstrated by Wagner and Grunewald[15] who obtained values of 7.2×10^{-9} and 7.5×10^{-9} $\mathrm{g^2\,cm^{-4}\,hr^{-1}}$ for oxygen partial pressures of 1 atm and 0.022 atm respectively. These results adequately confirm the conclusions drawn from a consideration of the defect structure of ZnO.

Systems forming multiple scale layers

Oxidation of iron

When iron oxidises in air at high temperature it grows a scale consisting of layers of FeO, Fe_3O_4, and Fe_2O_3 and thus provides a good example of the formation of multi-layered scales. Due to its importance in society, the oxidation of iron has been extensively investigated and consequently is relatively well understood with a very wide literature.

From the phase diagram of the iron–oxygen system, shown in Fig. 4.5, it is

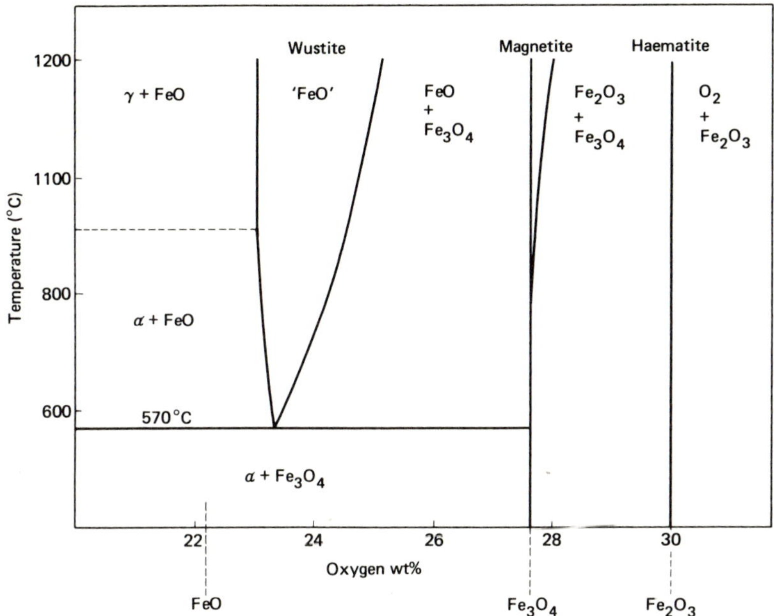

Fig. 4.5 The iron–oxygen phase diagram

clear that the phase wustite, FeO, does not form below 570 °C. Thus iron oxidised below this temperature would be expected to form a two-layered scale of Fe_3O_4 and Fe_2O_3 with the Fe_3O_4 next to the metal. Above 570 °C the oxide layer sequence in the scale would be FeO, Fe_3O_4, Fe_2O_3, with the FeO next to the metal.

The wustite phase, FeO, is a p-type metal deficit semiconductor which can exist over a wide range of stoichiometry, from $Fe_{0.95}O$ to $Fe_{0.88}O$ at 1000 °C according to Engell[16]. With such high cation vacancy concentrations, the mobility of cations and electrons (via vacancies and electron holes) is extremely high.

The phase magnetite, Fe_3O_4, is an inverse spinel, which therefore has divalent ions, Fe^{2+}, occupying octahedral sites and half of the trivalent ions, Fe^{3+}, occupying tetrahedral sites. Defects occur on both sites and consequently iron ions may diffuse over both tetrahedral and octahedral sites. A certain degree of intrinsic semiconduction is shown and consequently electrons may diffuse outwards over electron holes and, as excess electrons, in the conduction band. Except at high temperatures, only slight variation in stoichiometry is found.

Haematite, Fe_2O_3, exists in two forms, α-Fe_2O_3, which has a rhombohedral structure, and γ-Fe_2O_3, which is cubic. However, Fe_3O_4 oxidises to form α-Fe_2O_3 above 400 °C and only this structure need be considered[17]. In the rhombohedral crystal, the oxygen ions exist on a close packed hexagonal arrangement with iron ions in interstices. With such a structure it may be expected that iron ions would be mobile. However, α-Fe_2O_3 has been reported to show disorder on the anion sub-lattice only[18] from which only the oxygen ions are expected to be mobile. More recent results have suggested that Fe_2O_3 growth occurs by outward cation migration[19]. This aspect deserves further clarification and definitive investigation.

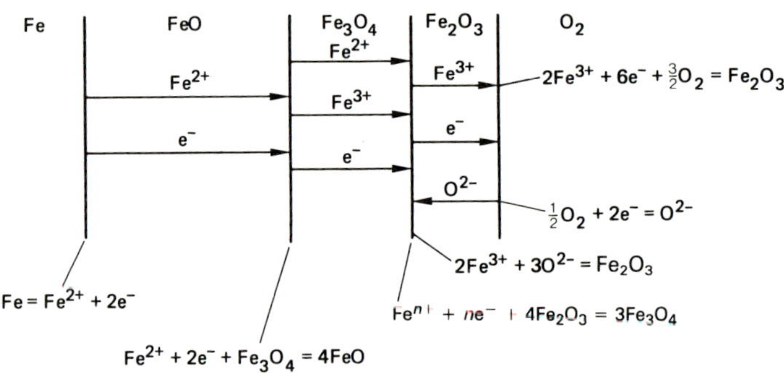

Fig. 4.6 Oxidation mechanism of iron to form a three-layered scale of FeO, Fe_3O_4, and Fe_2O_3 above 570 °C showing diffusion steps and interfacial reactions

On the basis of the above knowledge of the structure and diffusion properties of the iron oxides, a relatively simple mechanism can be proposed to represent the oxidation of iron as shown in Fig. 4.6. At the iron–wustite interface, iron ionises according to

$$Fe = Fe^{2+} + 2e^- \quad (4.10)$$

The iron ions and electrons migrate outwards through the FeO layer over iron vacancies and electron holes respectively. At the wustite–magnetite interface, magnetite is reduced by iron ions and electrons according to

$$Fe^{2+} + 2e^- + Fe_3O_4 = 4FeO \quad (4.11)$$

Iron ions and electrons surplus to this reaction proceed outward through the magnetite layer, over iron ion vacancies on the tetrahedral and octahedral sites and over electron holes and excess electrons respectively. At the magnetite–haematite interface, magnetite is formed according to

$$Fe^{n+} + ne^- + 4Fe_2O_3 = 3Fe_3O_4 \quad (4.12)$$

The value of n being 2 or 3 for Fe^{2+} or Fe^{3+} ions respectively.

If iron ions are mobile in the haematite they will migrate through this phase over iron ion vacancies V_{Fe}''' together with electrons and new haematite will form at the Fe_2O_3–gas interface according to

$$2Fe^{3+} + 6e^- + \tfrac{3}{2}O_2 = Fe_2O_3 \tag{4.13}$$

At this interface also, oxygen ionises according to

$$\tfrac{1}{2}O_2 + 2e^- = O^{2-} \tag{4.14}$$

If oxygen ions are mobile in the haematite layer, the iron ions and electrons, in excess of requirements for reduction of haematite to magnetite, will react with oxygen ions diffusing inwards through the Fe_2O_3 layer over oxygen vacancies forming new Fe_2O_3 according to

$$2Fe^{3+} + 3O^{2-} = Fe_2O_3 \tag{4.15}$$

The corresponding electrons then migrate outwards through the Fe_2O_3 to take part in the ionisation of oxygen at the Fe_2O_3–gas interface.

Due to the much greater mobility of defects in wustite, this layer will be very thick compared with the magnetite and haematite layers. In fact, the relative thicknesses of $FeO:Fe_3O_4:Fe_2O_3$ are in the ratio of roughly 95:4:1 at 1000 °C[20].

At temperatures below 570 °C, the wustite phase does not form and only the magnetite and haematite layers are seen in the scale. The rate of scaling is correspondingly low in the absence of wustite.

Due to the rapid rate of reaction of iron above 570 °C, thick scales are soon developed and, in spite of the relatively high plasticity of the FeO layer scale, metal adhesion is lost and a porous inner layer of FeO is formed, next to the metal, by the mechanism described earlier. The stresses associated with a rapidly growing scale undoubtedly induce physical defects in the outer scale and the penetration of gas molecules, especially those belonging to the CO/CO_2 and H_2/H_2O redox systems, will play a role in scale formation.

Since the scale formed on iron above 570 °C is predominantly wustite, growth of this layer controls the overall rate of oxidation. However, since the defect concentrations in wustite at the iron–wustite and wustite–magnetite interfaces are fixed by the equilibrium achieved there, for any given temperature, the parabolic rate constant will be relatively unaffected by the external oxygen partial pressure. Increasing the oxygen partial pressure in the gas phase should theoretically lead to an increase in the relative thickness of the haematite layer. However, since this layer only accounts for about 1% of the total scale thickness, any variation in rate constant with oxygen partial pressure will be difficult to detect.

A similar argument applies to temperatures below 570 °C in atmospheres of low oxygen partial pressure; low defect concentrations at the magnetite–iron and magnetite–haematite interfaces are fixed by the equilibria achieved there.

If it were possible to oxidise iron above 570 °C in atmospheres of low oxygen partial pressure, within the wustite existence range, under conditions leading to equilibrium being achieved at the wustite–gas interface, then a variation of parabolic rate constants with oxygen partial pressure should be observed. Unfortunately, the oxygen partial pressures required to demonstrate this are so low (10^{-12} atm at 1000 °C) that they can only be achieved using redox gas systems. Pettit, Yinger and Wagner used the CO/CO_2 system to investigate the oxidation of iron under such conditions and found that the rate was controlled by the decomposition of adsorbed CO_2 on the scale surface leading to a constant reaction rate as described earlier. For this reason, it is not possible to observe the variation in parabolic rate constant with oxygen partial pressure when iron is oxidised to give a scale composed of wustite only.

Oxidation of cobalt

Cobalt forms two oxides CoO and Co_3O_4 of NaCl and spinel structures respectively. CoO is a p-type cation deficient semiconductor through which cations and electrons migrate over cation vacancies and electron holes. In addition to the usual extrinsic defects due to deviations from stoichiometry, above 1050 °C intrinsic Frenkel type defects are also present[20]. The variations of oxidation rate constant with oxygen partial pressure and with temperature are, therefore, expected to be relatively complex. Consequently, it is important to ensure that very accurate data are obtained for the oxidation reactions, over a wide range of oxygen pressure and temperature.

The early data obtained up to 1966 are summarised in the available texts on high temperature oxidation of metals particularly those by Kofstad[21] and Mrowec and Werber[13].

In a very careful study, Mrowec and Przybylski[22] investigated the oxidation of cobalt between 940 and 1300 °C at oxygen pressures between 6.58×10^{-4} and 0.658 atm. They introduced refinements to the measurements such as

(a) Taking into account the thermal expansion of the metal which can increase the area at temperature by about 10% over the area measured at room temperature.
(b) Using flat, thin (19 mm × 15 mm × 0.5 mm) specimens so that surface area changes due to metal consumption during the reaction were held to about 3%.

Platinum markers, applied to the cobalt samples before oxidation, were found at the metal–scale interface after reaction over the whole range of oxygen and temperature investigated. This confirms that the compact CoO layer formed by outward diffusion of the metal.

Since the oxygen pressures used were orders of magnitude higher than the decomposition pressure of CoO the variation of the parabolic rate constant k_p'' with oxygen partial pressures

$$k_p'' = \text{const.}\ [p_{O_2}^{1/n} - (p_{O_2}^{0})^{1/n}] \tag{4.16}$$

can be written as

$$k_p'' = \text{const.}\ p_{O_2}^{1/n} \tag{4.17}$$

The rate constant will vary with temperature according to the Arrhenius equation, thus the variation of k_p'' with oxygen partial pressure and temperature can be written

$$k_p'' = \text{const}'.\ p_{O_2}^{1/n} \exp(-Q/RT) \tag{4.18}$$

Figure 4.7 shows Mrowec and Przybylski's results plotted accordingly.

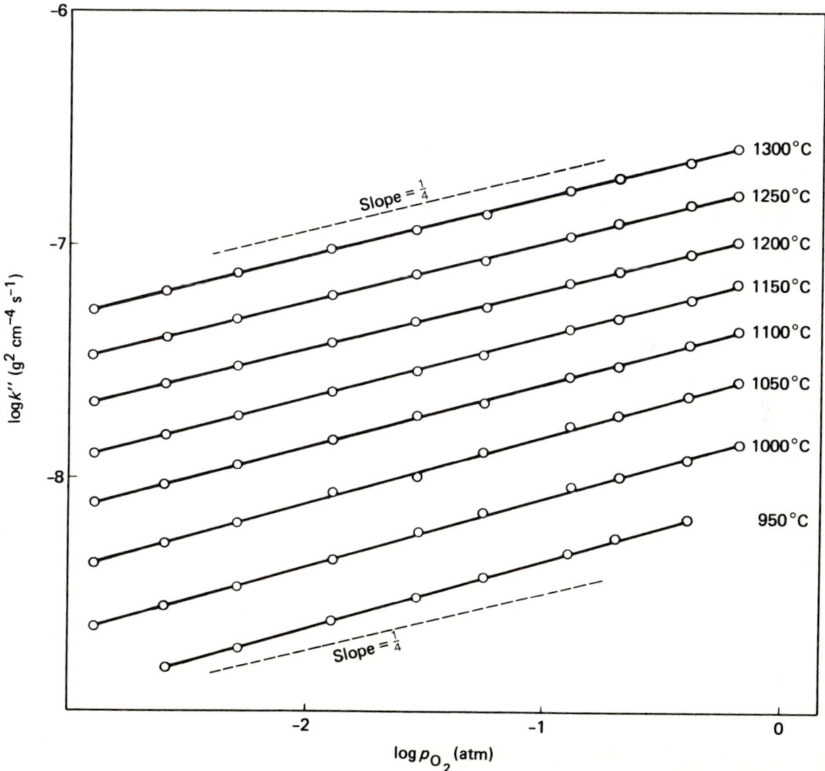

Fig. 4.7 Parabolic rate constant for the oxidation of cobalt to CoO at various oxygen partial pressures and temperatures, determined by Mrowec and Przybylski

From Fig. 4.7 it can be deduced that n varies from 3.4 at 950 °C to 3.96 at 1300 °C whereas Q increases from 159.6 kJ mol^{-1} at p_{O_2} = 0.658 atm to 174.7 kJ mol^{-1} at 6.58×10^{-4} atm. They further showed that these variations could be explained satisfactorily only by assuming that CoO contains defects arising intrinsically, i.e. Frenkel defects, as well as defects arising extrinsically as a result of deviations from stoichiometry.

The above work was carried out totally within the CoO range of existence. Earlier work by Bridges *et al.*[23] studied the oxidation of cobalt over a wider range of oxygen pressures well into the Co_3O_4 range of existence and over the temperature range 950–1150 °C. The results show much more scatter than those of Mrowec and Przybylski in the CoO field but also demonstrate clearly how the oxidation rate constant ceases to vary with oxygen pressure once the two-layer CoO–Co_3O_4 scale such as that in Fig. 4.8 is formed.

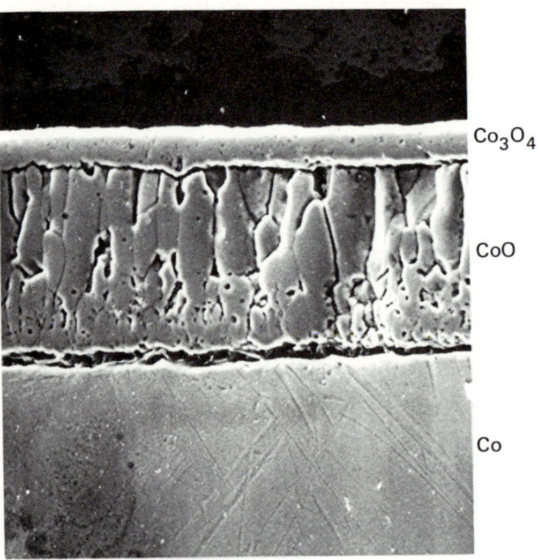

Fig. 4.8 Two-layered scale formed on cobalt after 10 hours in oxygen at 750 °C

A similar example to this is given for the case of copper[24] demonstrating clearly the dependence of the parabolic rate constant on oxygen partial pressure so long as the scale is a single layer of Cu_2O. As soon as an outer layer of CuO forms, the parabolic rate constant becomes independent of oxygen partial pressure.

Theory of multi-layered scale growth

As mentioned above, the formation of multiple scale layers is common for metals such as Fe, Co, and Cu as well as others.

The theory of multi-layered scale growth on pure metals has been treated by Yurek *et al.*[25]. The hypothetical system treated is shown in Fig. 4.9. It is assumed that the growth of both scales is diffusion controlled with the outward migration of cations large relative to the inward migration of anions. The flux of cations in each oxide is assumed to be independent of distance.

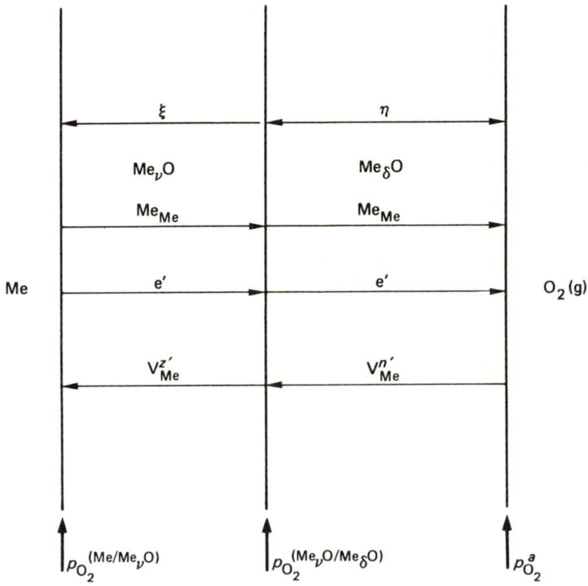

Fig. 4.9 Schematic diagram of a hypothetical two-layered scale

Each oxide exhibits predominantly electronic conductivity and local equilibrium exists at the phase boundaries. The total oxidation reaction is

$$(\nu w + \delta y)\text{Me} + \left(\frac{w+y}{2}\right)\text{O}_2 \rightarrow w\text{Me}_\nu\text{O} + y\text{Me}_\delta\text{O} \tag{4.19}$$

where w and y are the partition fractions of the two oxides in the scale.

The rate of thickening of a single product layer of $Me_\nu O$ would be

$$\frac{d\xi}{dt} = J_{\text{Me}}\left(\frac{V_{\text{Me}_\nu\text{O}}}{\nu}\right) \tag{4.20}$$

where J_{Me} is the outward flux of cations in moles cm^{-2} s^{-1}. However, in the case of two-layered scale growth only a fraction $w\nu/(\delta y + \nu w)$ of the cations transported through, $Me_\nu O$, leads to its growth with the remainder being transported further through the outer phase, $Me_\delta O$. Treatment of this case[25] yields

$$\frac{d\xi}{dt} = \frac{k_p(\text{Me}_\nu\text{O})}{\left(1 + \dfrac{\delta y}{\nu w}\right)} \tag{4.21}$$

where k_p is the parabolic rate constant for exclusive growth of $Me_\nu O$ and the bracketed term accounts for the partitioning of cations. A similar treatment of the growth of the outer layer yields

$$\frac{d\eta}{dt} = \frac{k_p(\mathrm{Me}_\delta\mathrm{O})}{\eta\left(1 + \frac{w}{y}\right)} \qquad (4.22)$$

Combinations of equations 4.21 and 4.22 allow calculation of the overall scale growth rate and integration of the two equations allows the thickness ratio of the two layers to be calculated.

This theory has been shown by Garnaud[26] to describe the growth of CuO and Cu_2O on Cu and by Garnaud and Rapp[27] to describe the growth of Fe_3O_4 and FeO on Fe.

Systems for which volatile species are important

Oxidation of chromium

The oxidation of pure Cr is, in principle, a simple process since a single oxide Cr_2O_3 is observed to form. However, under certain exposure conditions, several complications arise which are important both for the oxidation of pure Cr and for many important engineering alloys which rely on a protective Cr_2O_3 layer for oxidation protection. The two most important features are scale thinning by CrO_3 evaporation, and scale buckling as a result of compressive stress development.

The formation of CrO_3 by the reaction

$$Cr_2O_3(s) + \tfrac{3}{2}O_2(g) \rightleftharpoons 2CrO_3(g) \qquad (4.23)$$

becomes significant at high temperatures and high oxygen pressures. This was illustrated in Figs. 2.6 and 2.7. The evaporation of CrO_3, shown schematically in Fig. 4.10, results in the continuous thinning of the protective Cr_2O_3 scale so

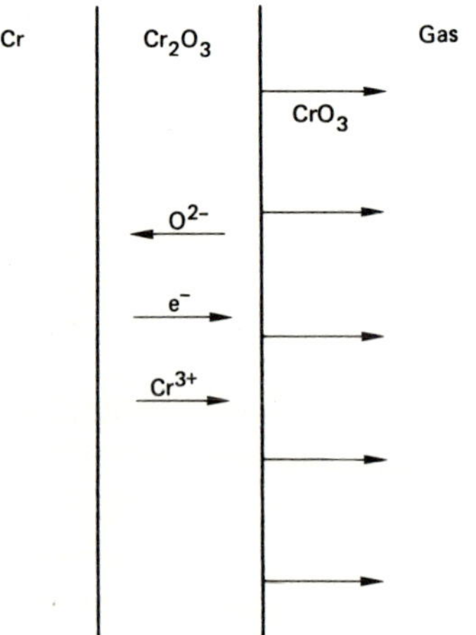

Fig. 4.10 Schematic diagram of combined scale growth and oxide volatilisation from Cr

the diffusive transport through it is rapid. The effect of the volatilisation on the oxidation kinetics has been analysed by Tedmon[28]. The instantaneous change in scale thickness is the sum of two contributions: thickening due to diffusion and thinning due to volatilisation.

$$\frac{dx}{dt} = \frac{k'_d}{x} - k'_s \tag{4.24}$$

where k'_d is a constant describing the diffusive process and k'_s describes the rate of volatilisation. This equation can be rearranged to yield

$$\frac{dx}{\left(\frac{k'_d}{x} - k'_s\right)} = dt \tag{4.25}$$

which, upon integration, yields:

$$\frac{-x}{k'_s} - \frac{k'_d}{k'^2_s} \ln(k'_d - k'_s x) + C = t \tag{4.26}$$

where C is an integration constant to be evaluated from the initial conditions. Taking $x = 0$ at $t = 0$

$$t = \frac{k'_d}{k'^2_s}\left[-\frac{k'_s}{k'_d}x - \ln\left(1 - \frac{k'_s}{k'_d}x\right)\right] \tag{4.27}$$

Initially, when the diffusion through a thin scale is rapid, the effect of CrO_3 volatilisation is not significant but, as the scale thickens, the rate of volatilisation becomes comparable and then equal to the rate of diffusive growth. This situation, paralinear oxidation, results in a limiting scale thickness, x_0, for which $dx/dt = 0$ which is shown schematically in Fig. 4.11. Setting this condition in equation 4.24 yields

$$x_0 = \frac{k'_d}{k'_s} \tag{4.28}$$

Since k'_d and k'_s have different activation energies, the value of x_0 will depend on temperature. The occurrence of a limiting scale thickness implies protective behaviour but, in fact, the amount of metal consumed increases until a constant rate is achieved. This can be seen more clearly by considering the metal recession, y, rather than the scale thickness.

$$\frac{dy}{dt} = \frac{k_d}{y} + k_s \tag{4.29}$$

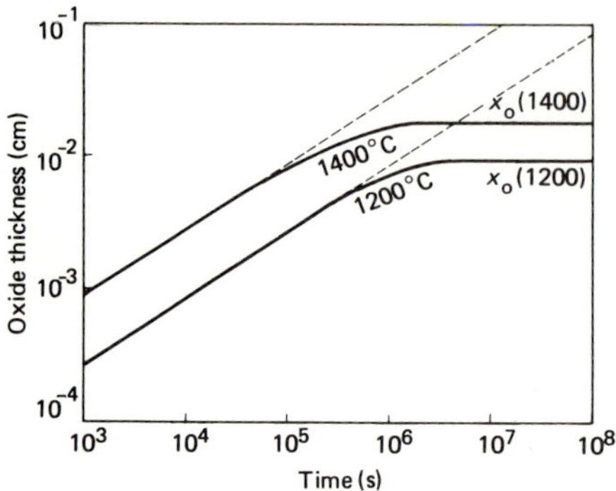

Fig. 4.11 Scale thickness versus time for the oxidation of Cr

Upon integration this yields

$$t = \frac{k_d}{k_s^2}\left[\frac{k_s}{k_d}y - \ln\left(1 + \frac{k_s}{k_d}y\right)\right] \tag{4.30}$$

When log y is plotted against log t, the curve shown in Fig. 4.12 is obtained.

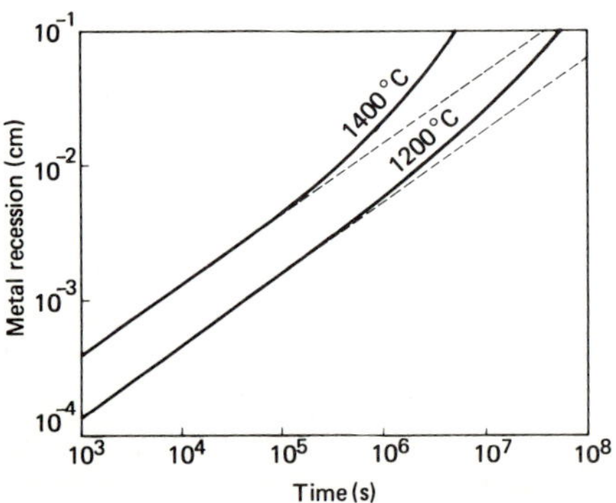

Fig. 4.12 Metal recession versus time for the oxidation of Cr

Here it is clear that the metal consumption is accelerating. This problem, which is more serious in rapidly flowing gases, is one of the major limitations on the high temperature use of Cr_2O_3-forming alloys and coatings.

The second important factor associated with the oxidation of Cr is the apparent variation of the oxidation rate with different surface preparations and the buckling of the Cr_2O_3 scales due to compressive stress development[29]. The latter observation is illustrated in Fig. 4.13. These results were somewhat puzzling until the careful work of Caplan and Sproule[30]. In this study, both electropolished and etched Cr single crystals were oxidised in 1 atm O_2 at 980,

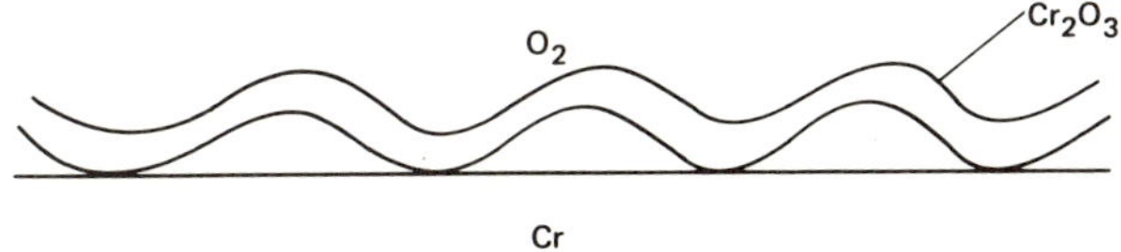

Fig. 4.13 Schematic diagram of scale buckling on Cr

1090, and 1200 °C. The electropolished specimens oxidised relatively rapidly and showed evidence of compressive stresses in the polycrystalline Cr_2O_3 scale which formed. The etched specimens showed similar behaviour for the scale formed in some areas, but on certain areas very thin single crystalline Cr_2O_3 was observed, showing no evidence of compressive stresses. Caplan and Sproule concluded that the single crystalline oxide grew by both outward transport of cations and inward transport of anions. The latter transport apparently occurs along oxide grain boundaries and results in compressive stresses by oxide formation at the scale–metal interface. While the transport through Cr_2O_3 needs further study, these results indicate the importance of grain boundary transport in high temperature oxidation.

Oxidation of molybdenum and tungsten

The volatilisation of oxides is particularly important in the oxidation of Mo and W at high temperatures and high oxygen pressures. Unlike Cr, which develops a limiting scale thickness, complete oxide volatilisation can occur in these systems. The condensed and vapour species for the Mo–O and W–O systems have been reviewed by Gulbransen and Meier[31] and the vapour species diagrams for a temperature of 1250 K are presented in Figs. 4.14 and 4.15. The effects of oxide volatility on the oxidation of Mo have been observed by Gulbransen and Wysong[32] at temperatures as low as 475 °C and the rate of oxide evaporation above 725 °C was such that gas phase diffusion became the rate controlling process[33]. Naturally, under these conditions, the rate of oxidation is catastrophic. Similar behaviour is observed for the oxidation of tungsten, but at higher temperatures because of the lower vapour pressures of the tungsten oxides[31]. The oxidation behaviour of tungsten has been reviewed in detail by Kofstad[21].

Oxidation of platinum

The oxidation of Pt, and Pt-group metals, is influenced by oxide volatility in that the only stable oxides are volatile. This results in a continuous mass loss. Alcock and Hooper[34] studied the mass loss of Pt and Rh at 1400 °C as a function of oxygen pressure. These results are presented in Fig. 4.16 where it is seen that the mass loss is directly proportional to the oxygen pressure. The gaseous species were identified as PtO_2 and RhO_2.

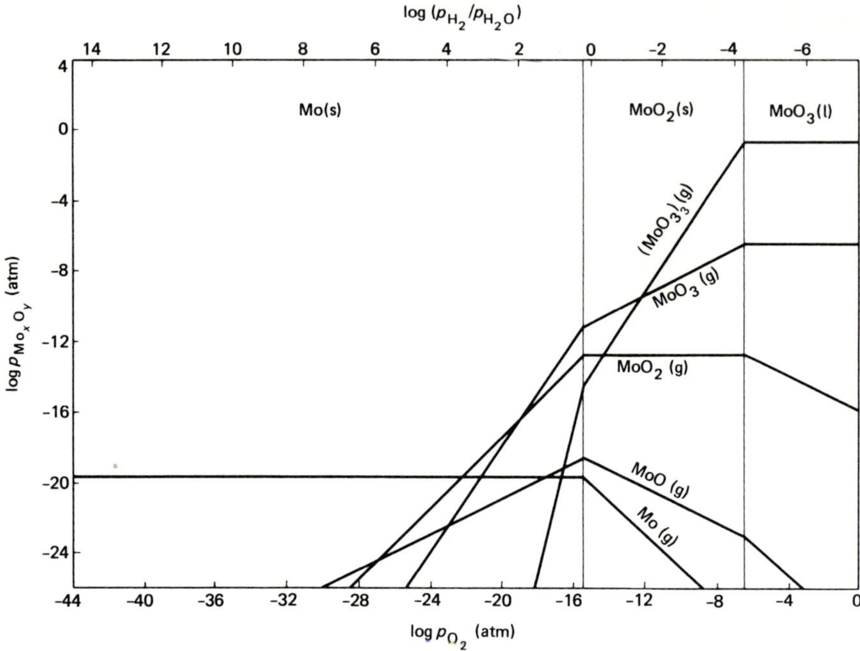

Fig. 4.14 Mo–O system volatile species at 1250 K

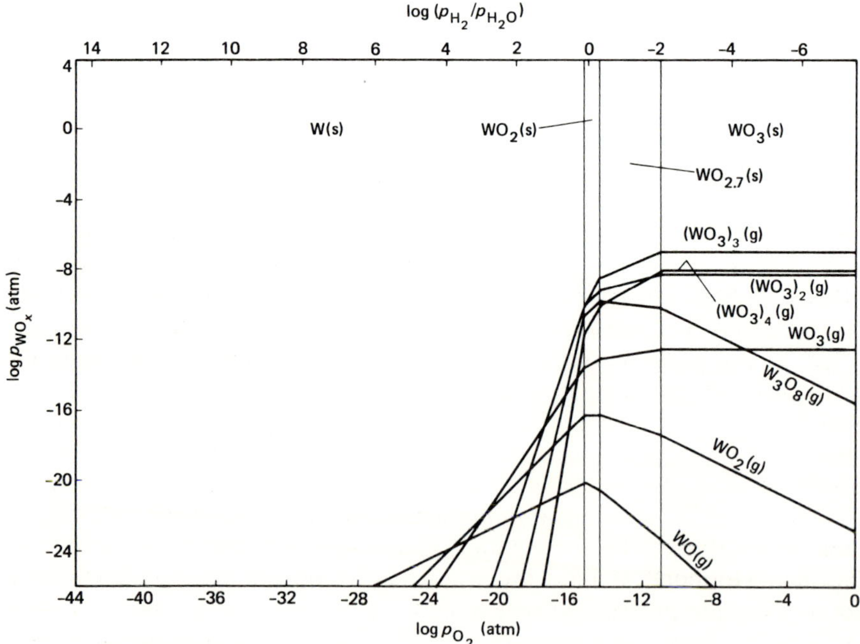

Fig. 4.15 Wo–O system volatile species at 1250 K

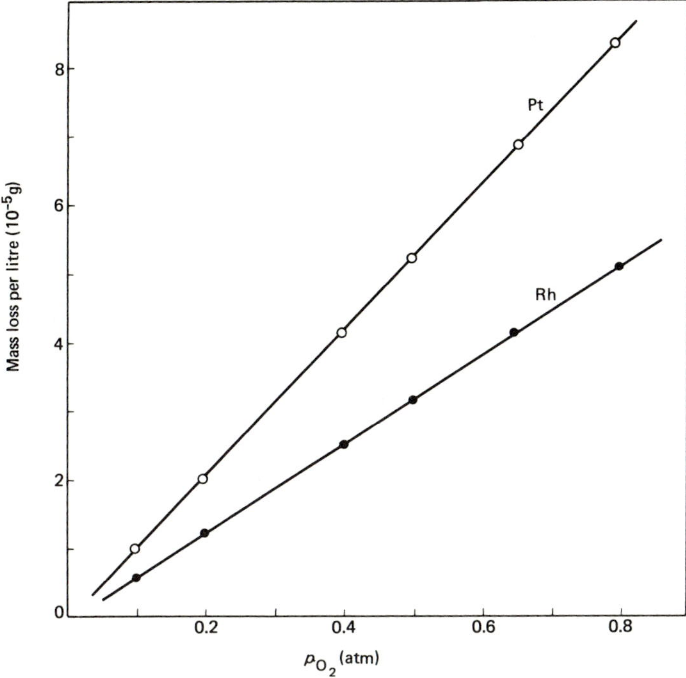

Fig. 4.16 Mass loss of gaseous Pt and Rh oxides as a function of p_{O_2} at 1400 °C (from Alcock and Hooper[34])

Oxidation of silicon

The formation of SiO_2 on silicon or silicon-containing alloys results in very slow oxidation rates. However, this system is also one which can be influenced markedly by oxide vapour species. Whereas the oxidation of Cr is influenced by such species at high oxygen pressures, the effects for Si are important at low oxygen pressures. The reason for this may be seen from the volatile species diagram for the Si–O system (see Fig. 4.17). A significant pressure of SiO is seen to be in equilibrium with SiO_2 (s) and Si (s) at oxygen pressures near the dissociation pressure of SiO_2. This can result in a rapid flux of SiO away from the specimen surface and the subsequent formation of a non-protective SiO_2 smoke. The formation of the SiO_2 as a smoke rather than as a continuous layer allows continued rapid reaction.

Wagner[35,36] has analysed the conditions under which this 'active' oxidation occurs. A simplified version[36] of this analysis follows. The fluxes of O_2 and SiO across a hydrodynamic boundary layer, taken to be the same thickness, δ, for both species, are given by

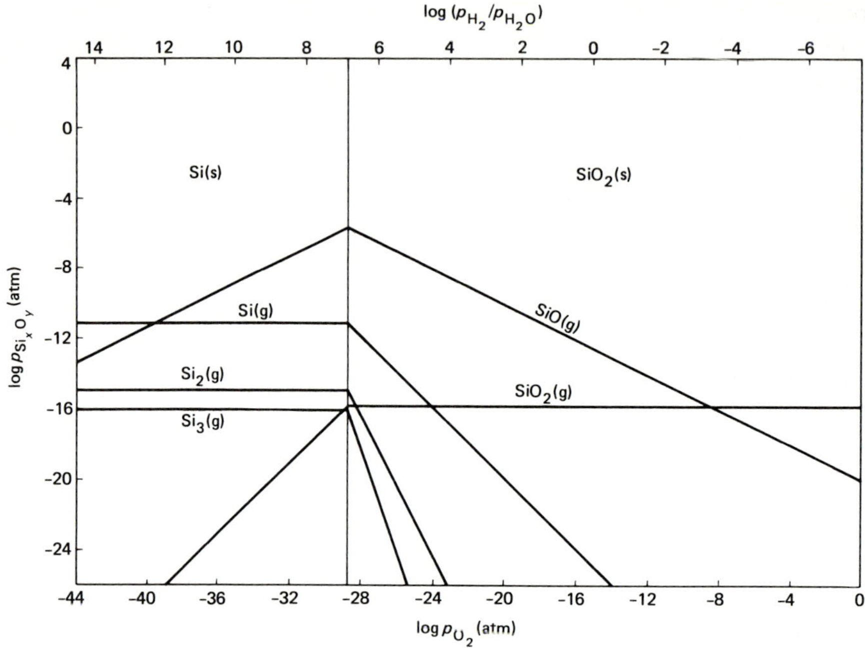

Fig. 4.17 Si–O system volatile species at 1250 K

$$J_{O_2} = \frac{p_{O_2} D_{O_2}}{\delta RT} \tag{4.31}$$

and

$$J_{SiO} = \frac{p_{SiO} D_{SiO}}{\delta RT} \tag{4.32}$$

where p_{SiO} is the pressure of SiO at the Si surface and p_{O_2} is the oxygen pressure in the bulk gas. Under steady state conditions, the net transport rate of oxygen atoms must vanish. Therefore

$$2J_{O_2} = J_{SiO} \tag{4.33}$$

which, upon substitution of equations 4.31 and 4.32 and assuming $D_{SiO} \approx D_{O_2}$, yields

$$p_{SiO} = 2p_{O_2} \tag{4.34}$$

If the equilibrium SiO pressure from the reaction

$$\tfrac{1}{2}\text{Si(s or l)} + \tfrac{1}{2}\text{SiO}_2\text{(s)} = \text{SiO(g)} \tag{4.35}$$

is greater than the SiO pressure in equation 4.34, the Si surface will remain bare and Si will be continually consumed. Therefore, there is a critical oxygen pressure

$$p_{O_2}(\text{crit}) \approx \tfrac{1}{2} p_{SiO}(\text{eq}) \tag{4.36}$$

below which 'active' oxidation will occur and the rate of metal consumption will be controlled by the SiO flux away from the surface.

$$J_{SiO} = \frac{2 p_{O_2} D_{O_2}}{\delta RT} \tag{4.37}$$

A more rigorous treatment of the problem[35] yields a slightly different expression for the critical oxygen pressure

$$p_{O_2}(\text{crit}) = \tfrac{1}{2} \left(\frac{D_{SiO}}{D_{O_2}} \right)^{1/2} p_{SiO}(\text{eq}) \tag{4.38}$$

Gulbransen *et al.*[37] have tested Wagner's prediction by oxidising Si over a range of temperatures and oxygen pressures and found good agreement. The rate of Si consumption was approximately 300 times faster for oxygen pressures below p_{O_2}(crit) than those above it. Interestingly, this is a phenomenon which is rare in high temperature oxidation but rather more common in aqueous corrosion, i.e. the rate of reaction being slower at higher driving forces (passive) than lower driving forces (active).

Systems with significant oxygen solubilities in the metal

Oxidation of titanium

The oxidation of Ti is quite complex because the Ti–O system exhibits a number of stable oxides and high oxygen solubility as seen from the phase diagram in Fig. 4.18. The rate laws observed for the oxidation of Ti vary with temperature as discussed by Kofstad[21]. However, in the temperature range 600 to 1000 °C, the oxidation is parabolic but the rate is a combination of two processes, oxide scale growth and oxide dissolution into the metal. The mass change per unit area is given by

$$\frac{\Delta m}{A} = k_p\,(\text{oxide growth})\, t^{1/2} + k_p\,(\text{dissolution})\, t^{1/2} \tag{4.39}$$

Similar behaviour is observed for the oxidation of Zr and Hf[21].

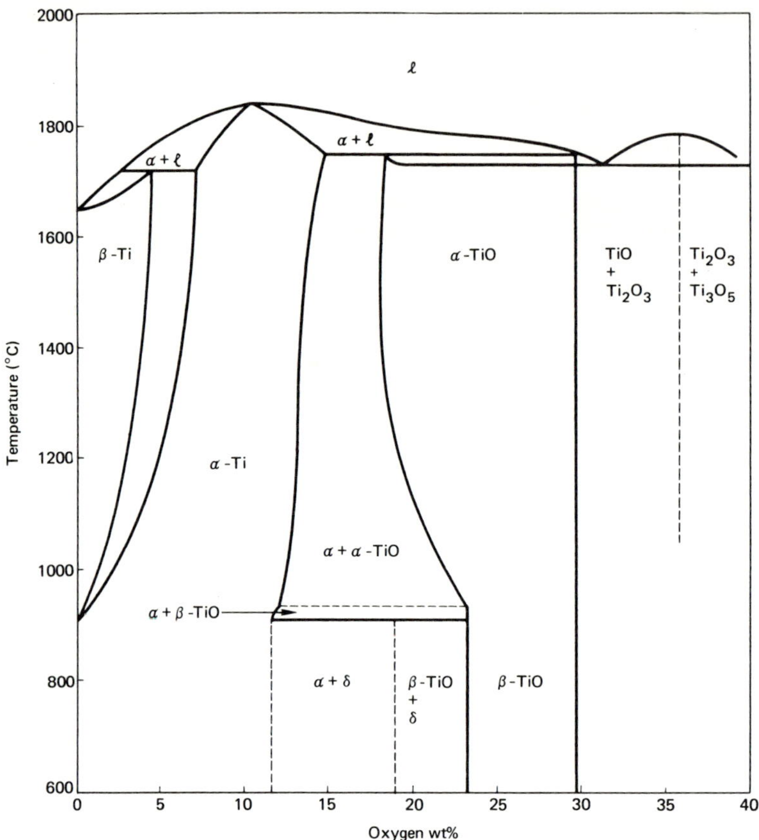

Fig. 4.18 The Ti-rich portion of the Ti–O phase diagram

Systems with significant scale cracking

Oxidation of niobium

The high temperature oxidation of Nb is characterised by inward diffusion of oxygen through the scale. Initially, a protective layer is formed but, as the scale grows, the formation of oxide at the scale–metal interface stresses the oxide resulting in scale cracking and a 'breakaway' linear oxidation. Roberson and Rapp[38] have shown this by an elegant experiment in which Nb was oxidised in sealed quartz tubes containing a mixture of Cu and Cu_2O powders at 1000 °C. The oxygen is transported from the powder by Cu_2O molecules and Cu is deposited at the reduction site. Figure 4.19 shows schematically the results of these experiments. Initially, the Cu was found at

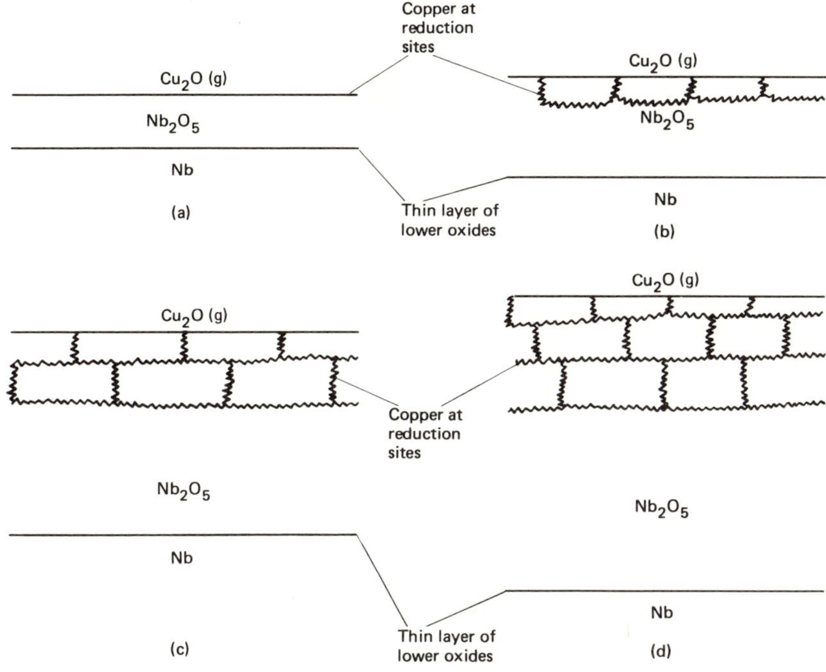

Fig. 4.19 Schematic diagram of the high temperature oxidation of Nb with oxygen supplied by a Cu/Cu_2O mixture (from Roberson and Rapp[38])

the scale–gas interface indicating the growth of a continuous layer of Nb_2O_5 over lower oxides of Nb by the inward migration of oxygen. With increasing time, the Cu was found to occupy cracks through the outer portion of the scale indicating that progressive scale cracking is the cause of the accelerated, linear, oxidation. Similar oxidation behaviour is also observed for Ta. As a result, neither Nb nor Ta can be used for any extended period at elevated temperature without a protective coating.

References

1 Romanski, J., *Corros. Sci.*, **8,** 67, 1968; also **8,** 89, 1968
2 Bruckmann, A., in: Reaction Kinetics in Heterogeneous Chemical Systems, ed. P. Barret, p. 409, *Proc. 25th Int. Meeting of Societie Chem. Physique*, Dijon, July, 1974; Elsevier, 1975
3 Mrowec, S. and Stoklosa, A., *Oxid. Metals*, **3,** 291, 1971
4 Kubaschewski, O. and van Goldbeck, O., *Z. Metallkunde*, **39,** 158, 1948
5 Matsunaga, Y., *Japan Nickel Rev.*, **1,** 347, 1933

6 Moore, W. J., *J. Chem. Phys.*, **19,** 255, 1951
7 Gulbransen, E. A. and Andrew, K. F., *J. Electrochem. Soc.*, **101,** 128, 1954
8 Philips, W., *J. Electrochem. Soc.*, **110,** 1014, 1963
9 Wood, G. C. and Wright, I. G., *Corros. Sci.*, **5,** 841, 1965
10 Birks, N. and Rickert, H., *J. Inst. Metals*, **91,** 308, 1962
11 Ilschner, B. and Pfeiffer, H., *Naturwissenschaft*, **40,** 603, 1953
12 Czerski, L. and Franik, F., *Arch. Gorn. Hutn.*, **3,** 43, 1955
13 Mrowec, S. and Werber, T., *Gas Corrosion of Metals*, U.S. Dept. of Commerce, Nat. Tech. Inf. Service, Springfield, VA 22161, 1978; p. 383
14 Fueki, K. and Wagner, J. B., *J. Electrochem. Soc.*, **112,** 384, 1965
15 Wagner, C. and Grunewald, K., *Z. Phys. Chem.*, **40B,** 455, 1938
16 Engell, H. J., *Acta Met.*, **6,** 439, 1968
17 Davies, M. H., Simnad, M. T. and Birchenall, C. E., *J. Metals*, **3,** 889, 1951
18 Bevan, D. J. M., Shelton, J. P. and Anderson, J. S., *J. Chem. Soc.*, 1729, 1948
19 Bruckman, A. and Simkovich, G., *Corros. Sci.*, **12,** 595, 1972
20 Mrowec, S. and Przybylski, K., *Oxid. Metals*, **11,** 383, 1977
21 Kofstad, P., *High Temperature Oxidation of Metals*, Wiley, New York, 1966
22 Mrowec, S. and Przybylski, K., *Oxid. Metals*, **11,** 365, 1977
23 Bridges, D. W., Baur, J. P. and Fassel, W. M., *J. Electrochem. Soc.*, **103,** 619, 1956
24 Mrowec, S. and Stoklosa, A., *Oxid. Metals*, **3,** 291, 1971
25 Yurek, G. J., Hirth, J. P. and Rapp, R. A., *Oxid. Metals*, **8,** 265, 1974
26 Garnaud, G., *Oxid. Metals*, **11,** 127, 1977
27 Garnaud, G. and Rapp, A., *Oxid. Metals*, **11,** 193, 1977
28 Tedmon, C. S., *J. Electrochem. Soc.*, **113,** 766, 1966
29 Caplan, D., Harvey, A. and Cohen, M., *Corros. Sci.*, **3,** 161, 1963
30 Caplan, D. and Sproule, G. I., *Oxid. Metals*, **9,** 459, 1975
31 Gulbransen, E. A. and Meier, G. H., Mechanisms of oxidation and hot corrosion of metals and alloys at temperatures of 1150 to 1450 K under flow, *Proc. of 10 Materials Research Symposium, National Bureau of Standards Special Publications 561*, 1979; p. 1639
32 Gulbransen, E. A. and Wysong, W. S., *TAIME*, **175,** 628, 1948
33 Gulbransen, E. A., Andrew, K. F. and Brassart, F. A., *J. Electrochem. Soc.*, **110,** 952, 1963
34 Alcock, C. B. and Hooper, G. W., *Proc. Roy. Soc.*, **254A,** 551, 1960
35 Wagner, C., *J. Appl. Phys.*, **29,** 1295, 1958
36 Wagner, C., *Corros. Sci.*, **5,** 751, 1965
37 Gulbransen, E. A., Andrew, K. F. and Brassart, F. A., *J. Electrochem. Soc.*, **113,** 834, 1966
38 Roberson, J. A. and Rapp, R. A., *TAIME*, **239,** 1327, 1967

5 Oxidation of alloys

Introduction

Many of the factors described for the oxidation of pure metals also apply to the oxidation of alloys. However, alloy oxidation is generally much more complex as a result of some, or all, of the following.

(a) The metals in the alloy will have different affinities for oxygen reflected by the different free energies of formation of the oxides.
(b) Ternary and higher oxides may be formed.
(c) A degree of solid solubility between the oxides may exist.
(d) The various metal ions will have different mobilities in the oxide phases.
(e) The various metals will have different diffusivities in the alloy.
(f) Dissolution of oxygen into the alloy may result in sub-surface precipitation of oxides of one or more alloying elements (internal oxidation).

The present chapter describes the major effects occurring in alloy oxidation and their relation to the above factors. No attempt has been made to provide a complete survey of the extensive literature on this subject but, rather, examples which illustrate the important fundamentals are presented. This will be done by first classifying the types of reactions which occur, then describing additional factors which have significant influences on the oxidation process, and finally describing the use of coatings for oxidation protection. Subsequent chapters will describe the oxidation of alloys in complex environments such as mixed gases and in the presence of liquid deposits. Previous reviews of alloy oxidation include those by Kubaschewski and Hopkins[1], Hauffe[2], Bénard[3], Pfeiffer and Thomas[4], Kofstad[5], Birchenall[6], and Mrowec and Werber[7].

Classification of reaction types

Wagner[8] has presented a lucid categorisation of simple types of alloy oxidation according to the reaction morphologies. The classification presented here is a modification of that given by Wagner in which alloys will be grouped as (a) noble parent with base alloying elements and (b) base parent with base alloying elements.

Noble parent with base alloying elements

This group includes alloy bases, such as Au, Ag, Pt, etc., which do not form stable oxides under normal conditions with alloying elements, such as Cu, Ni, Fe, Co, Cr, Al, Ti, In, Be, etc., which form stable oxides. However, at reduced oxygen pressures, those metals, such as Cu, Ni, and Co, which have only moderately stable oxides (see Chapter 2) can behave as noble parent metals.

A useful example of this alloy class is the Pt–Ni system. The important factors are a low solubility for oxygen in Pt and a small negative standard free energy of formation for NiO. When the alloy is first exposed to the oxidising atmosphere, oxygen begins to dissolve in the alloy. However, due to the low oxygen solubility and only moderate stability of NiO, the oxide cannot nucleate internally. This results in the outward migration of Ni to form a continuous NiO layer on the alloy surface. Since Pt does not enter into oxide formation it is inevitable that it will concentrate at the scale–metal interface and, correspondingly, Ni will be denuded there. Schematic concentration gradients are shown in Fig. 5.1. The reaction rate may be limited either by the

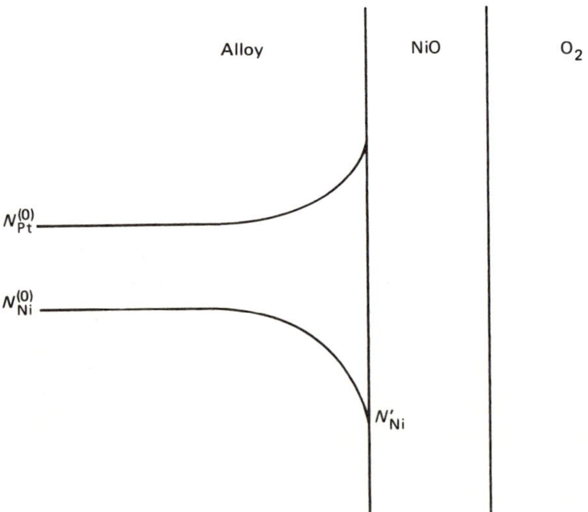

Fig. 5.1 Schematic concentration profiles for the oxidation of Pt–Ni alloys

supply of nickel from the alloy to the oxide or by the rate of diffusion of Ni ions across the oxide to the scale–gas interface where further oxide formation occurs. When the alloy nickel content is low, diffusion in the alloy is rate determining. When the nickel content is high, diffusion through the oxide is rate determining. This situation has been analysed by Wagner[9] and the predictions agree well with experiment. Both processes are diffusion controlled and both, therefore, lead to parabolic kinetics. As the nickel content is increased, the parabolic rate constant approaches values typical of pure Ni as shown schematically in Fig. 5.2.

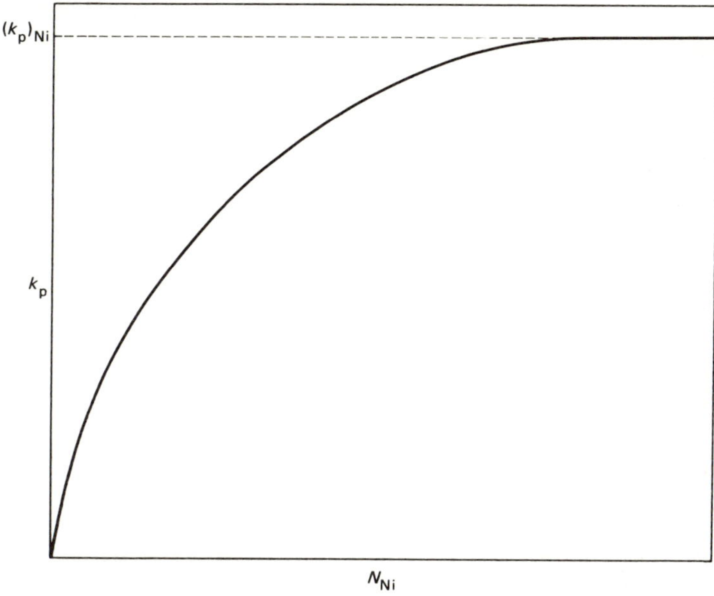

Fig. 5.2 Effect of nickel content on the parabolic rate constant for the oxidation of Pt–Ni alloys

When the scale growth rate is controlled by diffusion in the alloy, the scale–metal interface can become unstable as illustrated in Fig. 5.3 for a hypothetical alloy system A–B in which A is the noble element. Since any

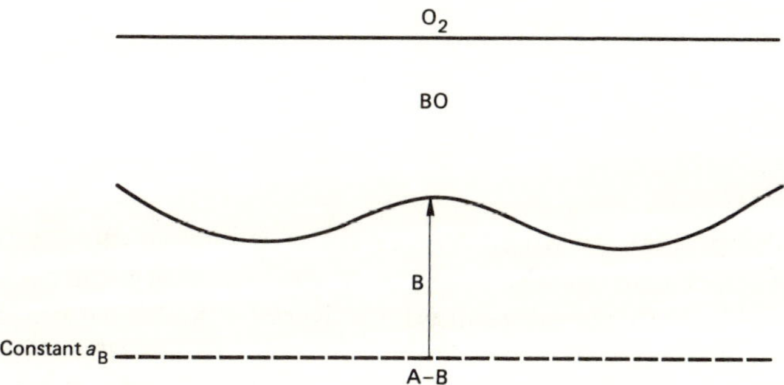

Fig. 5.3 Schematic representation of a non-planar scale–metal interface resulting from rate control by alloy interdiffusion

inward protuberance of the scale–metal interface will shorten the diffusion distance across the zone depleted in B, such a protuberance will grow and result in a wavy scale–metal interface. This problem has been described by Wagner[10]. Similar considerations apply to the scale–gas interface which can develop oxide whiskers if gas phase diffusion of oxidant to the surface controls the rate. Rickert[11] has shown this to be the case for the reaction of silver and sulphur using the experiment shown schematically in Fig. 5.4. A silver specimen was placed in one end of a sealed, evacuated tube and a quantity of sulphur was placed in the other. The tube was heated to 400 °C in an inclined furnace. Sulphur transport through the vapour caused a Ag_2S layer to grow on the portion of the silver nearest the sulphur. However, as the S was depleted from the vapour by this reaction, the rate of transport of S through the vapour to the far end of the specimen became rate controlling so that Ag_2S whiskers grew on this end as indicated in Fig. 5.4.

Silver-base alloys containing less noble solutes, such as In or Be, represent another useful case in this classification. In this case the solubility of oxygen in Ag is significant so that, for dilute alloys, internal oxidation can occur.

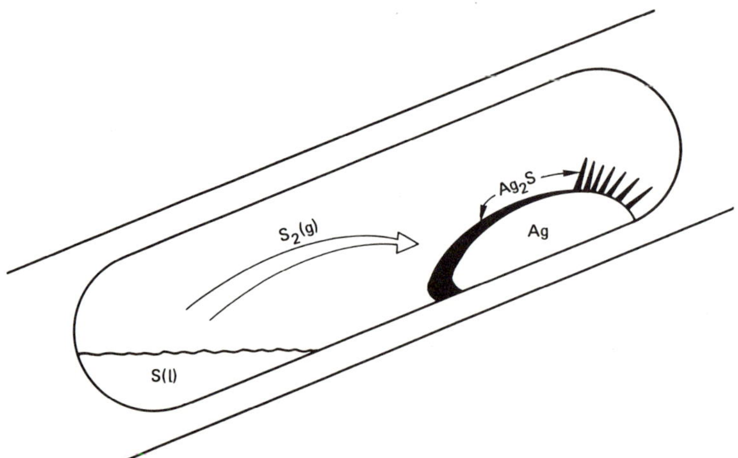

Fig. 5.4 Whisker growth in the reaction between Ag and S due to gas phase diffusion control

Internal oxidation

Internal oxidation is the process by which oxygen diffuses into an alloy and causes sub-surface precipitation of the oxides of one or more alloying elements. It has been the subject of reviews by Rapp[12], Swisher[13], and Meijering[14].

The necessary conditions for the occurrence of the phenomenon are

(a) $\Delta G^{\ominus}$ of formation (per mole O_2) for the solute metal oxide, BO_ν, must be more negative than $\Delta G^{\ominus}$ of formation (per mole O_2) for the base metal oxide.

(b) ΔG for the reaction

$$B + \nu\underline{O} = BO_\nu$$

must be negative. Therefore, the base metal must have a solubility and diffusivity for oxygen which is sufficient to establish the required activity of dissolved oxygen $\underline{O}$ at the reaction front.

(c) The solute concentration of the alloy must be lower than that required for the transition from internal to external oxidation.

(d) No surface layer must prevent the dissolution of oxygen into the alloy at the start of oxidation.

The process of internal oxidation occurs in the following manner. Oxygen dissolves in the base metal (either at the external surface of the specimen or at the alloy–scale interface if an external scale is present) and diffuses inward through the metal matrix containing previously precipitated oxide particles. The critical solubility product, $a_{\underline{B}}a_{\underline{O}}^{\nu}$, for the nucleation of precipitates is established at a reaction front (parallel to the specimen surface) by the inward diffusing oxygen and the outward diffusion of solute. Nucleation of the oxide precipitate occurs and a given precipitate grows until the reaction front moves forward and depletes the supply of solute B arriving at the precipitate. Subsequent growth occurs only by capillarity-driven coarsening (Ostwald ripening).

Kinetics

The rate of internal oxidation will be derived here for a planar specimen geometry using the quasi-steady state approximation. The results of similar derivations for cylindrical and spherical specimens will be presented. A more rigorous derivation of the kinetics of internal oxidation is presented in Appendix A.

Consider a planar specimen of a binary alloy A–B in which B is a dilute solute which forms a very stable oxide. Assume the ambient p_{O_2} is too low to oxidise A but high enough to oxidise B (see Fig. 5.5).

The quasi-steady state approximation assumes the dissolved oxygen concentration varies linearly across the zone of internal oxidation. Therefore, the oxygen flux through the internal oxidation zone (IOZ) is given by Fick's first law as

$$J = \frac{\mathrm{d}m}{\mathrm{d}t} = D_O \frac{N_O^{(S)}}{XV_m} \quad (\text{moles cm}^{-2}\,\text{s}^{-1}) \qquad (5.1)$$

where $N_O^{(S)}$ is the oxygen solubility in A (atom fraction), V_m is the molar volume of the solvent metal or alloy ($\text{cm}^3\,\text{mol}^{-1}$), and D_O is the diffusivity of

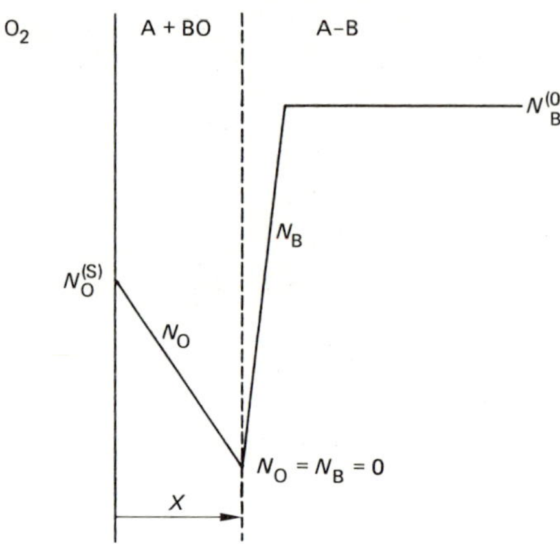

Fig. 5.5 Simplified concentration profiles for the internal oxidation of A–B

oxygen in A ($cm^2\ s^{-1}$). If counter-diffusion of solute B is assumed to be negligible, the amount of oxygen accumulated in the internal oxidation zone per unit area of reaction front is

$$m = \frac{N_B^{(0)} \nu X}{V_m} \quad (\text{moles cm}^{-2}) \tag{5.2}$$

where $N_B^{(0)}$ is the initial solute concentration. Differentiation of equation 5.2 with respect to time gives an alternate expression for the flux.

$$\frac{dm}{dt} = \frac{N_B^{(0)} \nu}{V_m} \frac{dX}{dt} \tag{5.3}$$

Equating 5.1 and 5.3

$$D_O \frac{N_O^{(S)}}{X V_m} = \frac{N_B^{(0)} \nu}{V_m} \frac{dX}{dt} \tag{5.4}$$

and rearranging equation 5.4 gives

$$X\,dX = \frac{N_O^{(S)} D_O}{\nu N_B^{(0)}}\,dt \tag{5.5}$$

Integration of equation 5.5, assuming $X = 0$ at $t = 0$, yields

$$\tfrac{1}{2}X^2 = \frac{N_O^{(S)} D_O}{\nu N_B^{(0)}}\,t \tag{5.6}$$

or

$$X = \left[\frac{2N_O^{(S)} D_O}{\nu N_B^{(0)}}\,t\right]^{1/2} \tag{5.7}$$

Equation 5.7 gives the penetration depth of the internal oxidation zone as a function of oxidation time. The following points concerning equation 5.7 are of interest.

(a) The penetration depth has a parabolic time dependence.

$$X \propto t^{1/2}$$

(b) The penetration depth for a fixed time is inversely proportional to the square root of the atom fraction of solute in the bulk alloy.
(c) Careful measurements of the front penetration as a function of time for an alloy of known solute concentration can yield a value for the solubility–diffusivity product, $N_O^{(S)} D_O$ (permeability) for oxygen in the matrix metal.

The results of a similar derivation for a cylindrical specimen[13] yield

$$\frac{(r_1)^2}{2} - (r_2)^2 \ln\left(\frac{r_1}{r_2}\right) + \tfrac{1}{2} = \frac{2N_O^{(S)} D_O}{\nu N_B^{(0)}} t \tag{5.8}$$

and for a spherical specimen[13]

$$\frac{(r_1)^2}{3} - (r_2)^2 + \tfrac{2}{3}\frac{(r_2)^3}{(r_1)} = \frac{2N_O^{(S)} D_O}{\nu N_B^{(0)}} t \tag{5.9}$$

where r_1 is the specimen radius and r_2 is the radius of the unoxidised alloy core.

Figure 5.6 shows a polished and etched cross-section of an internally oxidised Cu–Ti cylinder showing the penetration of the internal oxidation zone[15]. The TiO_2 particles are too small to resolve in the optical microscope. Measurements of r_2 from such cross-sections for various oxidation times allow the rate of penetration to be measured. Figure 5.7 shows the results of such measurements for Cu–Ti alloys at 850 °C. The curvature of the plot is predicted by equation 5.8. Plotting the left hand side of equation 5.8 versus time yields a straight line the slope of which is $2N_O^{(S)} D_O/\nu N_B^{(0)}$ from which $N_O^{(S)} D_O$ may be calculated[15].

Precipitate morphology

Many of the effects of internal oxidation, both on the overall corrosion process for an alloy and on influencing the mechanical, magnetic, etc., properties of the alloy, are intimately related to the morphology of the oxide precipitates. The following is a rather qualitative discussion of the factors which influence the size and shape of internal oxides.

In general, the oxide particle size is determined by a competition between the rate of nucleation as the internal oxidation front passes and the subsequent growth and coarsening rates of the particles. The longer the time a nucleated particle has to grow between the arrival of oxygen and solute at its

Fig. 5.6 Microstructure of a Cu–O.47 wt% Ti alloy internally oxidised at 800 °C for 97 hours

surface and the next nucleation event which depletes the supply of solute, the larger will be the particle. Subsequent coarsening may occur, but this phenomenon will be temporarily neglected. Therefore, those factors which favour high nucleation rates will decrease the particle size and those which favour high growth rates will increase the size.

If the nucleation of precipitates is assumed to be controlled by the velocity

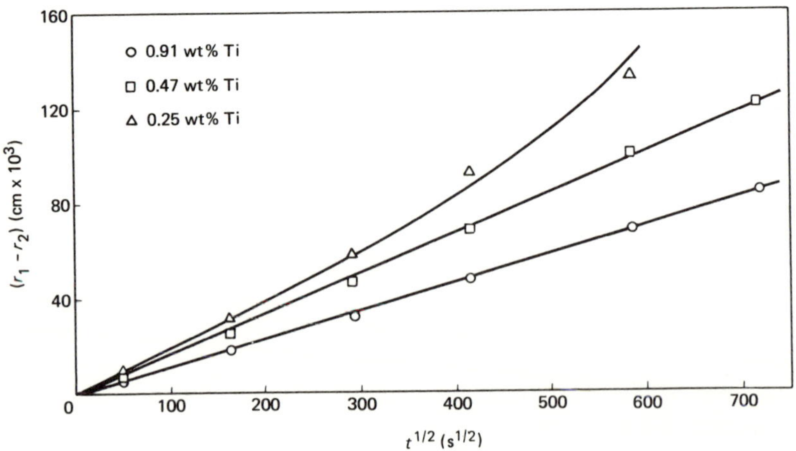

Fig. 5.7 Internal oxidation zone penetration as a function of time at 850 °C

of the oxidation front (i.e. the rate at which a critical supersaturation may be built up) and the growth is controlled by the length of time available for growth of a precipitate, the size should be inversely proportional to the front velocity, i.e. $r \propto 1/v$. Rewriting equation 5.5 gives the velocity for a planar specimen as

$$v = \frac{dX}{dt} = \frac{N_O^{(S)} D_O}{\nu N_B^{(0)}} \frac{1}{X} \tag{5.10}$$

Therefore, other things being equal, one would predict the particle size to increase proportionally to (a) X, (b) $N_B^{(0)}$, and (c) $1/N_O^{(S)} \propto (1/p_{O_2})^{1/2}$.

The effect of temperature is difficult to evaluate but, apparently, the nucleation rate is affected much less by increasing temperature than is the growth rate so that particle size increases with increasing temperature of internal oxidation. As the front penetration increases, the velocity eventually decreases to a point where diffusion of solute to a growing particle occurs rapidly enough to prevent nucleation of a new particle. This results in the growth of elongated or needle-like particles[16], e.g. Fig. 5.8.

The effect of particle–matrix interfacial free energy is often overlooked but is particularly important in the nucleation and coarsening of internal oxides. Consider the classical nucleation problem of forming a spherical nucleus. The free energy change for the formation of a nucleus of radius r is

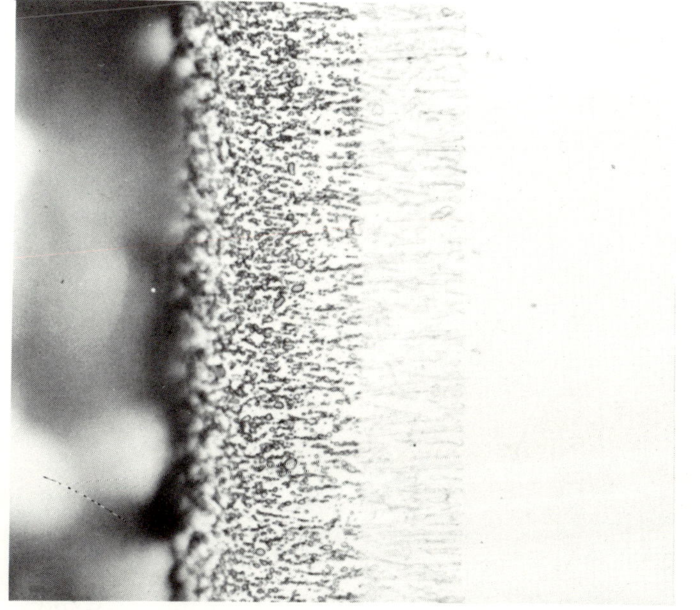

Fig. 5.8 Co–5 wt% Ti alloy internally oxidised for 528 hours at 900 °C. Unetched

$$\Delta G = 4\pi r^2 \sigma + \frac{4}{3}\pi r^3 \Delta G_v \tag{5.11}$$

where σ is the specific interfacial free energy and ΔG_v is the free energy change per unit volume of precipitate or particles formed during the reaction. The surface and volume terms along with ΔG are plotted as a function of r in Fig. 5.9. Nuclei with radii greater than r^* will tend to grow spontaneously. At r^*

$$\frac{d\Delta G}{dr} = 8\pi r^* + 4\pi r^{*2} \Delta G_v = 0 \tag{5.12}$$

which may be solved for r^* as

$$r^* = \frac{-2\sigma}{\Delta G_v} \tag{5.13}$$

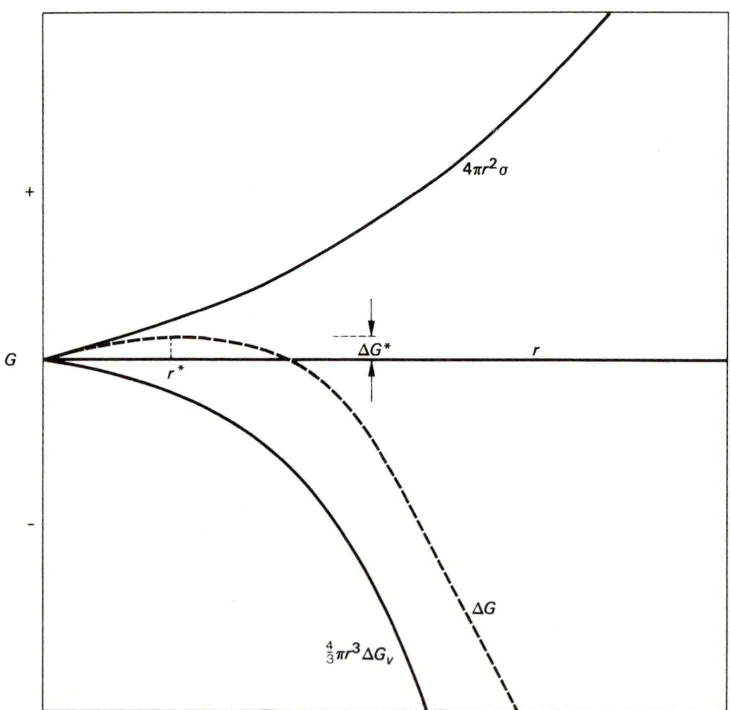

Fig. 5.9 Schematic plot of free energy versus radius of nucleating particle

Substitution into equation 5.11 yields ΔG^*, the height of the activation barrier to nucleation

$$\Delta G^* = 4\pi \frac{4\sigma^3}{\Delta G_v^2} - \frac{4}{3}\pi \frac{8\sigma^3}{\Delta G_v^3} \Delta G_v = \left(16 - \frac{32}{3}\right) \pi \frac{\sigma^3}{\Delta G_v^2} = \frac{16}{3}\pi \frac{\sigma^3}{\Delta G_v^2} \tag{5.14}$$

It is this value (ΔG^*) which enters into the nucleation rate as calculated from absolute reaction rate theory

$$J = \nu C^* = \nu C_{\mathrm{O}}\, \mathrm{e}^{-\Delta G^*/RT} \tag{5.15}$$

Clearly, since the nucleation rate is proportional to the exponential of a quantity, which in turn depends on the cube of the surface free energy, this parameter will have a drastic effect on the nucleation rate and particle size will increase with increasing σ. This influence will be felt further in the coarsening of the particles since the interfacial free energy is the driving force for this process.

In a similar manner the relative stability of the internal oxide, $\Delta G^{\ominus}$, will influence ΔG_v and $N_{\mathrm{O}}^{(\mathrm{S})}$ with more stable oxides resulting in higher nucleation rates and smaller particles.

In summary, the size of internal oxides will be determined by a number of factors with larger particles forming for

(a) deeper front penetrations, X,
(b) higher solute concentrations, $N_{\mathrm{B}}^{(0)}$,
(c) lower oxygen solubilities and lower p_{O_2},
(d) higher temperature,
(e) higher particle–matrix interfacial free energies,
(f) less stable oxides.

These predictions have been born out in the limited number of studies of particle morphology available in the literature[15–18].

Transition from internal to external oxidation

Consideration of equation 5.10 shows that the penetration velocity of the internal oxidation zone decreases with increasing $N_{\mathrm{B}}^{(0)}$ and decreasing $N_{\mathrm{O}}^{(\mathrm{S})}$ (i.e. decreased p_{O_2}). Thus there will be a limiting concentration of solute in the alloy above which the outward diffusion of B will be large enough to form a continuous blocking layer of BO_ν and stop internal oxidation. This transition to external oxidation is the basis for the design of Fe-, Ni-, and Co-base engineering alloys which contain a sufficiently high concentration of a solute (e.g. Cr, Al, or Si) to produce an external layer of a stable, slowly growing oxide (e.g. Cr_2O_3, Al_2O_3, or SiO_2) which prevents oxidation of the parent metal. This process, known as *selective oxidation*, will be discussed further but at this point the fundamentals of the transition from internal to external oxidation must be considered. The analysis of this

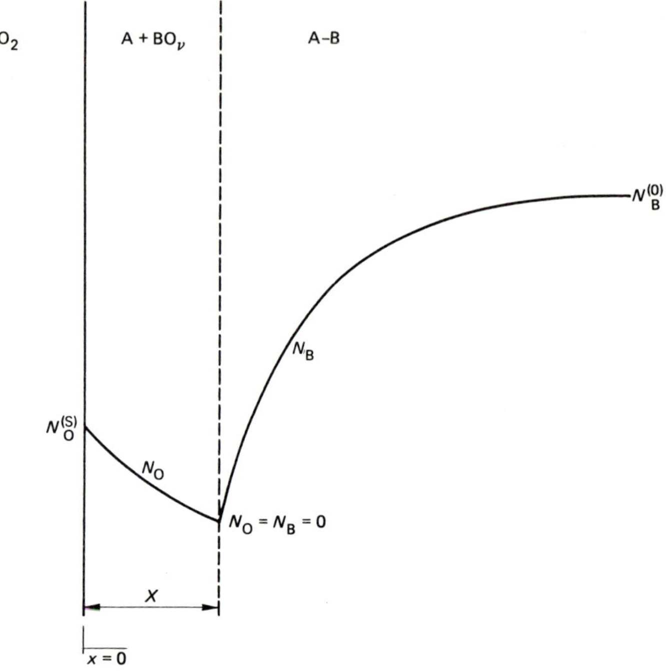

Fig. 5.10 Concentration profiles for internal oxidation of A–B

problem has been carried out by Wagner[19]. Consider the concentration profiles in Fig. 5.10. The penetration of the zone of internal oxidation may be expressed (see Appendix A)

$$X = 2\gamma (D_O t)^{1/2} \tag{5.16}$$

Fick's second law for the diffusion of oxygen through the internally oxidised zone is

$$\frac{\partial N_O}{\partial t} = D_O \frac{\partial^2 N_O}{\partial x^2} \tag{5.17}$$

with the initial conditions

$$\begin{aligned} t = 0 \quad N_O &= N_O^{(S)} \quad \text{for } x \leq 0 \\ N_O &= 0 \qquad \text{for } x > 0 \end{aligned}$$

and the boundary condition

$$t > 0 \quad N_O = N_O^{(S)} \quad \text{for } x = 0$$

The solution of equation 5.17 (following Appendix A) is

$$N_O(x,t) = N_O^{(S)}\left[1 - \frac{\operatorname{erf}\left(\dfrac{x}{2(D_O t)^{1/2}}\right)}{\operatorname{erf}\gamma}\right] \tag{5.18}$$

Similarly, Fick's second law for the counter-diffusion of B is

$$\frac{\partial N_B}{\partial t} = D_B\frac{\partial^2 N_B}{\partial x^2} \tag{5.19}$$

with initial conditions

$$t = 0 \quad N_B = 0 \quad \text{for } x < 0$$
$$N_B = N_B^{(0)} \quad \text{for } x > 0$$

The solution to equation 5.19, therefore, becomes

$$N_B(x,t) = N_B^{(0)}\left[1 - \frac{\operatorname{erfc}\left(\dfrac{x}{2(D_B t)^{1/2}}\right)}{\operatorname{erfc}(\theta^{1/2}\gamma)}\right] \tag{5.20}$$

where $\theta = D_O/D_B$.

If f represents the mole fraction of BO_ν in the internal oxidation zone and V_m the molar volume of BO_ν, then f/V_m will be the concentration in moles/volume and the number of moles in a volume element, $A\,dX$, will be $(f/V_m)A\,dX$, where A is the cross-sectional area for diffusion. This quantity must be equal to the number of moles of B arriving at $x = X$ in the time dt due to diffusion from within the sample, i.e. for $x > X$. Therefore

$$\frac{fA\,dX}{V_m} = \lim_{\varepsilon\to 0}\left[\frac{A\,D_B}{V_m}\frac{\partial N_B}{\partial x}\right]_{x = X+\varepsilon} dt \tag{5.21}$$

Substitution of equations 5.16 and 5.20 yields an enrichment factor of α, where

$$\alpha = \frac{f}{N_B^{(0)}} = \frac{1}{\gamma\pi^{1/2}}\left(\frac{D_B}{D_O}\right)^{1/2}\frac{\exp(-\gamma^2\theta)}{\operatorname{erfc}(\gamma\theta^{1/2})} \tag{5.22}$$

For the case of interest here, $D_O N_O^{(S)} \ll D_B N_B^{(0)}$, i.e. the oxygen permeation is significantly less than that of B, the product $\gamma\theta^{1/2} \ll 1$. Therefore

$$\alpha \approx \frac{2\nu}{\pi}\left[\frac{N_B^{(0)}\,D_B}{N_O^{(S)}\,D_O}\right] \tag{5.23}$$

When α is large, we expect the accumulation and lateral growth of the internal oxides to form a continuous layer, i.e. the transition to external oxidation. Figure 5.11 shows this occurring in a Co–7.5% Ti alloy[16]. Wagner states that when the volume fraction of oxide, $g = f(V_{ox}/V_m)$, reaches a critical value, g^*, the transition from internal to external scale formation

Fig. 5.11 Transition from internal to external oxidation in Co–7.5 wt% Ti alloy internally oxidised for 528 hours at 900 °C. Unetched

should occur. Rearrangement of equation 5.23 then gives the criterion for external oxidation as

$$N_{\mathrm{B}}^{(0)} > \left[\frac{\pi g^*}{2\nu} N_{\mathrm{O}}^{(\mathrm{S})} \frac{D_{\mathrm{O}} V_{\mathrm{m}}}{D_{\mathrm{B}} V_{\mathrm{ox}}}\right] \tag{5.24}$$

Equation 5.24 enables prediction of how changing exposure conditions will affect the concentration of solute required to produce an external scale. Those conditions which decrease the inward flux of oxygen, e.g. lower $N_{\mathrm{O}}^{(\mathrm{S})}$ (i.e. lower p_{O_2}) and those which increase the outward flux of B, e.g. cold working the alloy (increasing D_{B} by increasing the contribution of short-circuit diffusion) will allow the transition to external oxidation at lower solute concentrations. Rapp[20] has tested the Wagner theory for Ag–In alloys. It was found that at $p_{\mathrm{O}_2} = 1$ atm and $T = 550\,°\mathrm{C}$ the transition occurred for $g^* = 0.30$ ($N_{\mathrm{In}} = 0.16$). Rapp then used this value to predict the effect of reduced p_{O_2} (reduced $N_{\mathrm{O}}^{(\mathrm{S})}$) on the In concentration required for the transition. The agreement between experiment and equation 5.24 was excellent. Rapp's experiments also gave an indirect verification of the effect of increased D_{B} due to cold work on the transition. Specimens containing 6.8% In were observed to undergo the transition to external $\mathrm{In_2O_3}$ formation around scribe marks under exposure conditions which required 15% In to produce the transition on undeformed surfaces.

An additional factor which can have a pronounced influence on the transition is the presence of a second solute whose oxide has a stability intermediate between that of A and B. If the p_{O_2} is great enough to form the oxide of the second solute this will, in effect, decrease the inward flux of oxygen and allow external formation of BO_ν at lower solute concentrations[21]. This concept was verified by Wagner for Cu–Zn–Al alloys[21] and by Pickering for Ag–Zn–Al alloys[22]. In the latter experiments Ag–3 at% Al alloys were found to oxidise internally while under the same conditions a Ag–21 at% Zn–3 at% Al alloy produced an external Al_2O_3 scale. This concept proves useful in the design of high temperature alloys since the element which provides protection by selective oxidation (e.g. Al) often has detrimental effects on other alloy properties, particularly mechanical properties, and should be held at as low a concentration as is feasible. Thus, by addition of substantial concentrations of Zn, external scales of Al_2O_3 can be promoted.

The above systems have been considered since they allow some of the features of alloy oxidation to be examined without the complication of the simultaneous formation of two oxides. In principle, any alloy system can be examined in this way, provided the ambient oxygen partial pressure is below that required to form the lowest oxide of the more noble component.

Base parent with base alloying elements

This classification represents the systems most commonly used and the formation of two or more oxides must be considered. The principles of oxidation of this class of alloys will be developed through specific examples.

Cu–Be alloys

The oxidation of dilute Cu–Be alloys in oxygen or air results in internal oxidation of the solute to form BeO particles and external oxidation of the Cu matrix to form Cu_2O which, in some cases, is covered with a layer of CuO. The BeO particles do not react with the Cu_2O scale to form ternary oxides. Maak[23] has analysed the kinetics of this process which is illustrated in Fig. 5.12 for a hypothetical alloy A–B which forms oxides AO and BO_ν. The positions of the scale–metal interface and the internal oxidation front at time t are Y and X respectively, both measured from the original alloy surface which is taken as the origin for x. Clearly, equations 5.17 and 5.19 apply to this problem with the conditions for equation 5.17

$$N_O = 0 \quad \text{for } x > X,\ t > 0$$
$$N_O = N_O^{(S)} \quad \text{for } x = X,\ t > 0$$

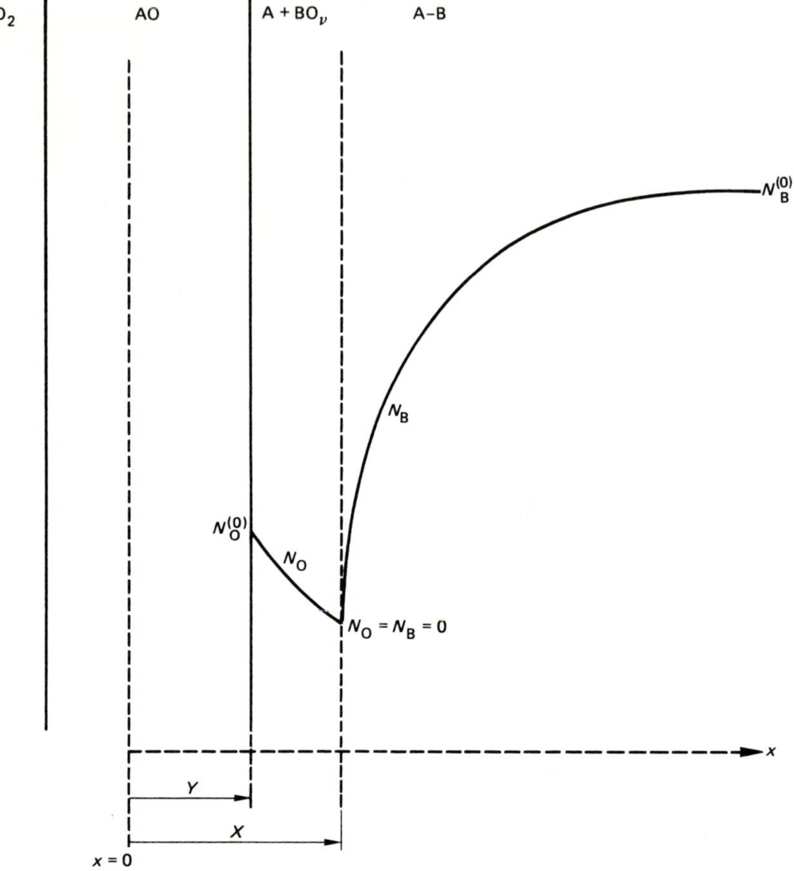

Fig. 5.12 Schematic diagram of the concentration profiles for the oxidation of Cu–Be-type alloys

and for equation 5.19

$$N_B = 0 \quad \text{for } x \leqslant X,\ t < 0$$
$$N_B = N_B^{(0)} \quad \text{for } x \geqslant 0,\ t = 0$$

Assuming parabolic oxidation of A we have

$$Y = (2k_c t)^{1/2} \tag{5.25}$$

and from Appendix A we assume

$$X = 2\gamma(D_O t)^{1/2} \tag{5.26}$$

The solution to equation 5.17 for this case is

$$N_O(x,t) = N_O^{(S)}\left[\frac{\operatorname{erf}\gamma - \operatorname{erf}\left(\dfrac{x}{2(D_O t)^{1/2}}\right)}{\operatorname{erf}\gamma - \operatorname{erf}\left(\dfrac{k_c}{2D_O}\right)^{1/2}}\right] \quad (5.27)$$

and the solution to equation 5.19 is

$$N_B(x,t) = N_B^{(0)}\left[1 - \frac{\operatorname{erfc}\left(\dfrac{x}{2(D_B t)^{1/2}}\right)}{\operatorname{erfc}(\gamma\theta^{1/2})}\right] \quad (5.28)$$

Since the stoichiometry of the internal oxide BO_ν requires equivalent fluxes of solute and oxygen at the internal oxidation front, we have

$$\lim_{\varepsilon \to 0}\left[-D_O\left(\frac{\partial N_O}{\partial x}\right)_{x = X - \varepsilon} = \nu D_B\left(\frac{\partial N_B}{\partial x}\right)_{x = X + \varepsilon}\right] \quad (5.29)$$

Substitution of equations 5.27 and 5.28 into equation 5.29 yields

$$\frac{N_O^{(S)} D_O^{1/2}\exp(-\gamma^2)}{\operatorname{erf}\gamma - \operatorname{erf}\left(\dfrac{k_c}{2D_O}\right)^{1/2}} = \frac{\nu N_B^{(0)} D_B^{1/2}\exp(-\gamma^2\theta)}{\operatorname{erfc}(\gamma\theta^{1/2})} \quad (5.30)$$

We generally have $N_O^{(S)} \ll N_B^{(0)}$, $\gamma \ll 1$, and $Y/2(D_O t)^{1/2} = (k_c/2D_O)^{1/2} \ll 1$ which allows equation 5.30 to be rearranged to give

$$X(X - Y) = \frac{N_O^{(S)} D_O}{\gamma N_B^{(0)}} 2t \frac{1}{\pi^{1/2}(\gamma\theta^{1/2})\exp(\gamma^{1/2}\theta)\operatorname{erfc}(\gamma\theta^{1/2})} \quad (5.31)$$

Careful measurement of X and Y as functions of time allows equation 5.31 to be used to calculate the oxygen permeability in A but these results are generally somewhat inaccurate due to porosity at the scale–metal interface. Maak's treatment has also been extended to describe the simultaneous motion of two internal oxidation fronts into an alloy[16].

Ni–Cr alloys

Extensive studies of the oxidation behaviour of Ni–Cr alloys have been published by a number of investigators[24,25]. Alloys in this system with low Cr contents show internal oxidation of Cr forming Cr_2O_3 islands within a matrix of almost pure Ni. An outer scale of NiO is formed with an inner layer, sometimes porous, of NiO containing $NiCr_2O_4$ islands as shown in Fig. 5.13.

The inner, duplex, layer consists of NiO with some Cr in solution which provides additional cation vacancies, thus increasing the mobility of Ni ions in this region. This doping effect, explained in Appendix B, results in an increase in the observed oxidation rate constant, compared with pure Ni, which is also increased due to the extra oxygen tied-up in Cr_2O_3 in the internal oxidation zone.

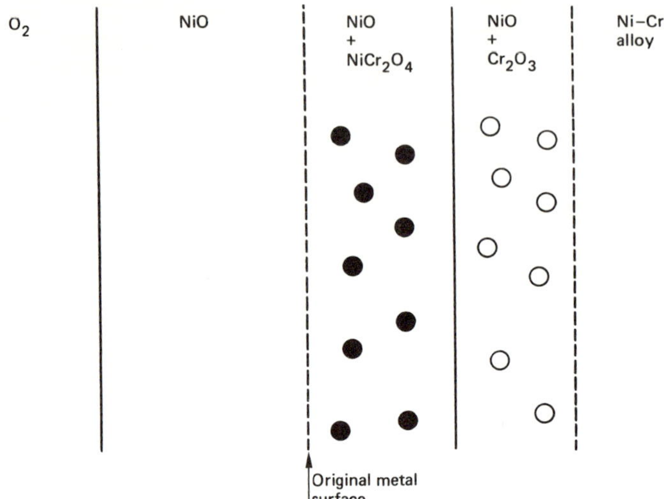

Fig. 5.13 Schematic diagram of the oxidation morphology of dilute Ni–Cr alloys

As the outer scale advances into the metal, the Cr_2O_3 internal oxide islands are engulfed by NiO and a solid state reaction ensues to form $NiCr_2O_4$

$$NiO + Cr_2O_3 = NiCr_2O_4 \tag{5.32}$$

These $NiCr_2O_4$ islands remain as a second phase in the scale in the cases of all but the most dilute Cr alloys and their extent marks the position of the original metal surface.

Since cation diffusion is much slower through the $NiCr_2O_4$ spinel than it is through NiO, the spinel islands in the scale act as diffusion blocks for outward migrating Ni ions. Consequently, as the Cr content of the alloy is increased, the increasing volume fraction of spinel reduces the total Ni flux through the scale and the rate constant begins to fall. Eventually as the Cr content is increased further, to about 10% at 1000 °C, the mode of oxidation changes, as with Ag–In alloys, to give a complete external scale of Cr_2O_3. At this and higher compositions the oxidation rate falls abruptly to values more typical of Cr than of Ni. This is illustrated in Fig. 5.14.

The above discussion of the effect of Cr content applies to steady state oxidation. In fact, when a clean alloy surface, even one with a Cr concentration in excess of the critical value, is exposed to an oxidising atmosphere, oxides containing Ni as well as Cr will form. Since the Ni-containing oxides grow much more rapidly than Cr_2O_3 a significant amount of NiO and $NiCr_2O_4$ can form before a continuous Cr_2O_3 layer can form. This phenomenon, known as *transient oxidation*, occurs for almost all alloy systems for which the oxides of more than one component have negative free energy changes for their formation in the given atmosphere. The nature of the transient oxidation

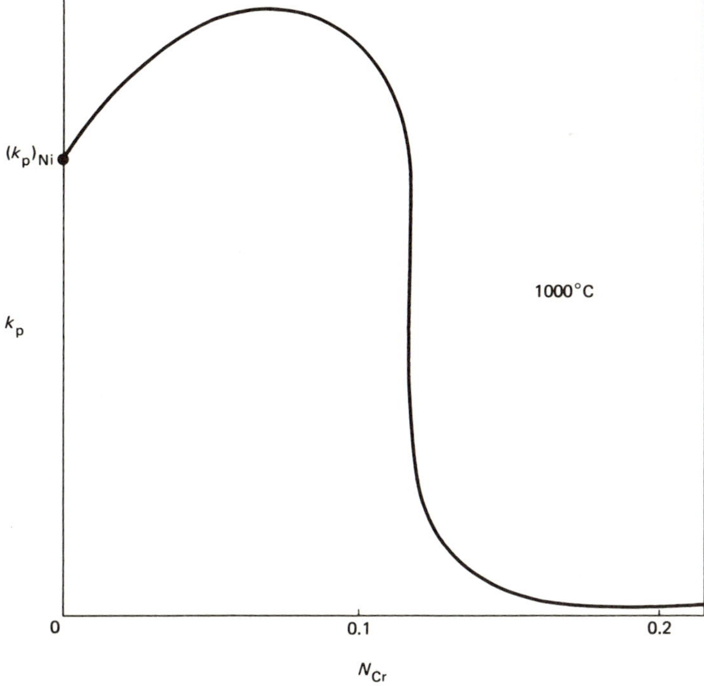

Fig. 5.14 Dependence of the parabolic rate constant for the oxidation of Ni–Cr alloys on Cr content

often affects the ultimate selective oxidation process in which only the most stable oxide is formed. The extent of the transient period is decreased by those factors which promote selective oxidation. For example, for Ni–Cr alloys the amount of Ni-rich oxide is observed to be less for higher alloy Cr contents, reduced oxygen pressures, and exposure of cold worked alloys.

With Cr contents of around the critical value, the protection conferred by the scale of Cr_2O_3 may not be permanent, since the selective oxidation of Cr to form the scale may denude the alloy below the scale of Cr to the point where its concentration is below the critical one. Therefore, any rupturing, or mechanical damage, of the scale will expose a lower Cr alloy which will undergo internal oxidation and NiO scale formation with a corresponding increase in reaction rate as illustrated in Fig. 5.15. In essence the system is returned to the transient oxidation period.

The phenomenon of depletion of an element being selectively oxidised is a common one already mentioned for Pt–Ni alloys and is illustrated for Ni–50% Cr alloys in Fig. 5.16. Since these alloys contain two phases, a Ni-rich fcc matrix and Cr-rich bcc precipitate, the Cr depletion is delineated by the depth to which the Cr-rich second phase has dissolved. The extent of

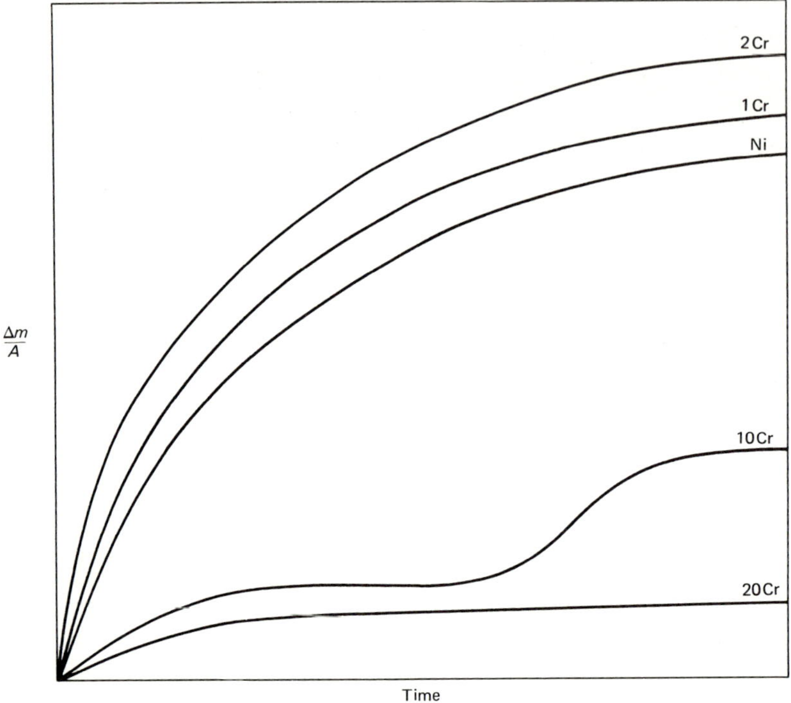

Fig. 5.15 Mass change versus time for Ni with various Cr contents

depletion of an element being selectively oxidised depends on a number of factors; its concentration, the rate of scale growth, and the alloy interdiffusion coefficient being the most important. Because of this depletion effect, oxidation resistant alloys based on the Ni–Cr system usually contain at least 18–20% Cr.

Fe–Cr alloys

As before, it is convenient to consider this system in order of increasing Cr content. At low Cr contents, no internal oxidation zone is seen. This is because the rate at which the external scale is formed is so rapid that the thickness of the internal oxidation zone is negligibly small. The system is best described in terms of the Fe–Cr–O phase diagram shown schematically in Fig. 5.17. The rhombohedral oxides Fe_2O_3 and Cr_2O_3 show a continuous series of solid solutions. The Fe- and Cr-oxides react to form spinels which form solid solutions with Fe_3O_4.

At low Cr contents, both chromium-rich and iron oxides form on the surface. Some Cr will enter solution in the FeO phase but, due to the stability of spinel, the solubility is limited as in the case of Ni–Cr alloys (see Fig. 5.18).

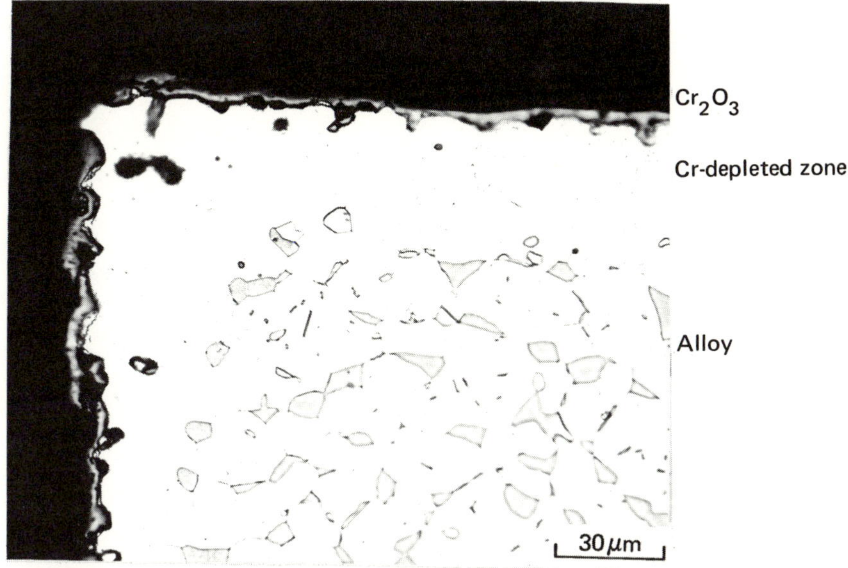

Fig. 5.16 Chromium-depleted zone for a Ni–50Cr alloy oxidised for 21 hours at 1100 °C in oxygen

Furthermore, the effect of a few more vacancies is not observable in the highly defective p-type FeO. Thus, an increase in rate constant is rarely observed. On increasing the Cr content, Fe^{2+} ions are progressively blocked by the $FeCr_2O_4$ islands and the FeO layer correspondingly becomes thinner relative to the Fe_3O_4 layer thickness. In these stages the reaction rate is still quite fast and typical of pure iron. When the Cr content is increased further, a scale of mixed spinel $Fe(Fe,Cr)_2O_4$ is produced and the parabolic rate constant is lower. Apparently, iron ions are much more mobile in this oxide than Cr^{3+} ions since at longer times quite pure iron oxides can be found at the outer surface of the scale when the oxidation rate is controlled by the diffusion of iron ions through the inner mixed-spinel layer.

On exceeding the Cr concentration N^*_{Cr} (Fig. 5.17) an outer scale is formed initially of Cr_2O_3 with a corresponding reduction in the parabolic rate constant. A permanent protective scale cannot be achieved unless N^*_{Cr} is exceeded and most Fe–Cr systems designed for heat resistance have in excess of 20% Cr. It should be emphasised that the popular 'stainless steels' 8% Ni–18% Cr were developed for corrosion resistance to aqueous environments only and should not be regarded as oxidation resistant at high temperatures – a common error. In fact, the Fe–Cr system is not a good system to use as the basis of high temperature oxidation resistant alloys for long-term exposures because of the range of solid solutions formed by the

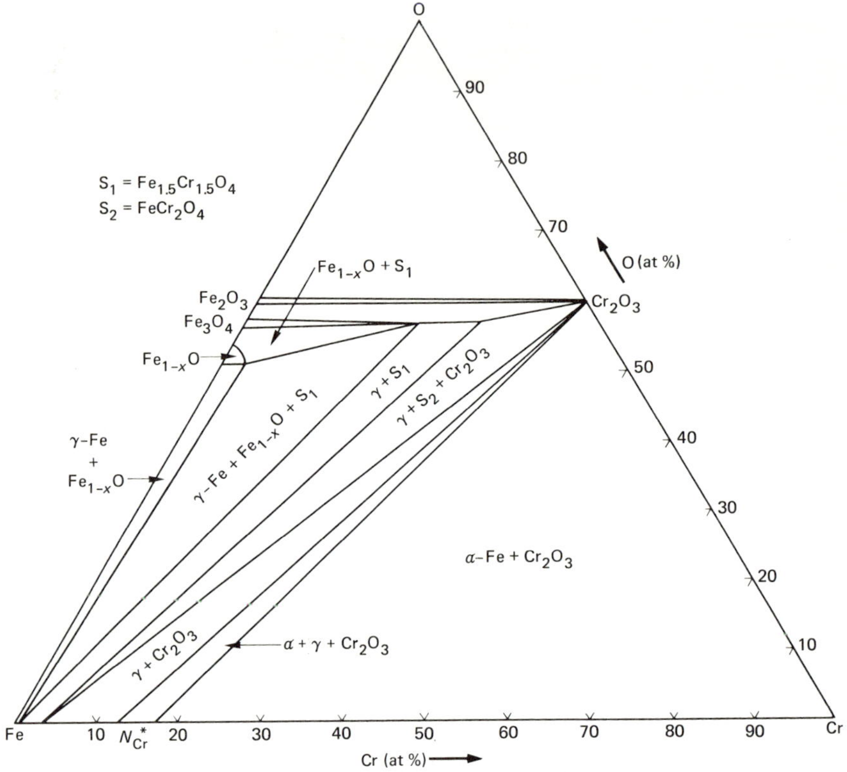

Fig. 5.17 Isothermal section of the Fe–Cr–O phase diagram at 1200 °C

Fe_2O_3–Cr_2O_3 rhombohedral oxides. Even with high Cr levels, iron ions will dissolve in and diffuse rapidly through the Cr_2O_3 scale and eventually an outer layer of fairly pure Fe oxides will result. This is illustrated in Fig. 5.19 for Fe–25% Cr which has been oxidised at 1150 °C for 24 hours. Protrusions of Fe_3O_4 (A) are seen projecting through the Cr_2O_3 scale (B). The use of Cr contents higher than 20–25% cannot be achieved without other alloying additions to avoid sigma phase formation with its attendant embrittlement of the alloy.

Fe–Si alloys

As with Fe–Cr alloys, this system does not generally show internal oxidation for the same reasons. The oxides involved in this system include SiO_2 which forms a silicate Fe_2SiO_4 with FeO, Fe_3O_4, and Fe_2O_3.

At low silicon contents, SiO_2 is formed at the alloy surface. It is apparently distributed very finely since, instead of being surrounded by the advancing

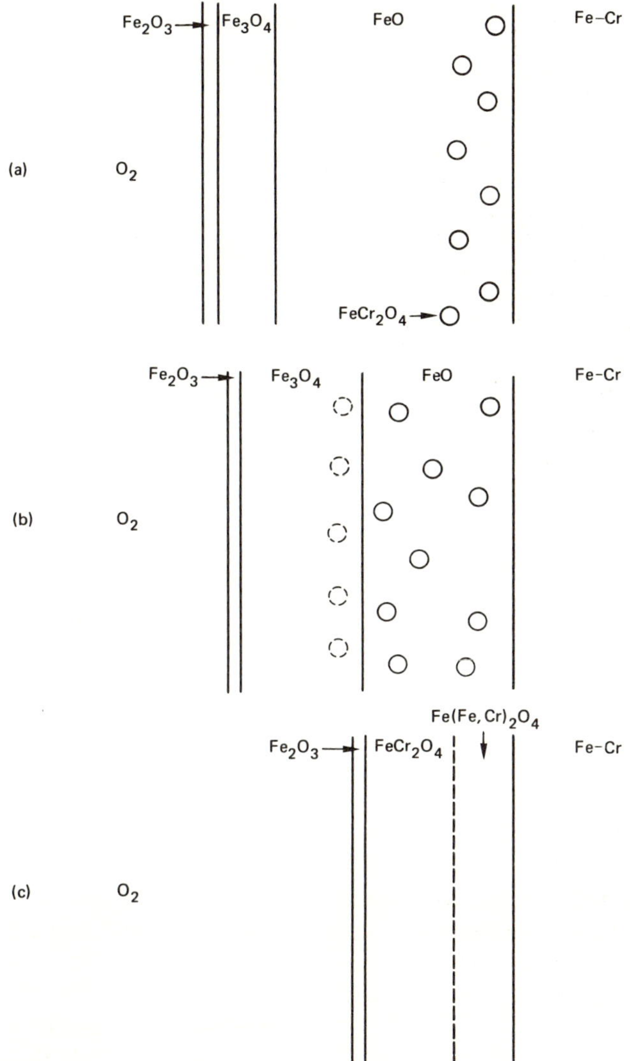

Fig. 5.18 Schematic diagrams of the scale morphologies on (a) Fe–5Cr, (b) Fe–10Cr, and (c) Fe–15Cr

iron oxide scale, it is swept along and accumulated at the alloy surface. Simultaneously it reacts with FeO to form fayalite, Fe_2SiO_4, and when the particles become larger they are engulfed by the scale. These islands lie in the FeO layer as markers but do not give an accurate indication of the position of the original metal–scale interface. The fayalite is indeed frequently seen to lie as stringers parallel to the metal surface (see Fig. 5.20).

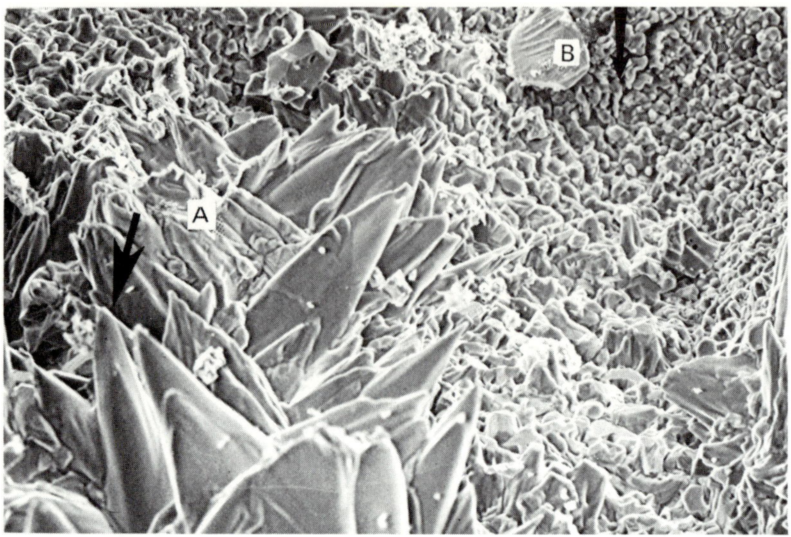

Fig. 5.19 Scale–gas interface of an Fe–25Cr alloy oxidised for 24 hours at 1150 °C

The silicon content can be increased to provide a continuous, protective SiO_2 scale. However, at the concentrations required, the formation of intermetallic compounds make the alloys unstable mechanically.

Both Fe–Cr and Fe–Si alloys can be made to show internal oxidation by reducing or preventing the inward movement of the outer scale. This can be demonstrated by oxidising in air to form a scale, then encapsulating the sample under vacuum. Alternatively, internal oxidation can be observed at areas where scale–metal separation has occurred, i.e. usually at corners and edges.

In the systems discussed so far, the alloying element has been much more reactive towards oxygen than the parent metal and has consequently tended

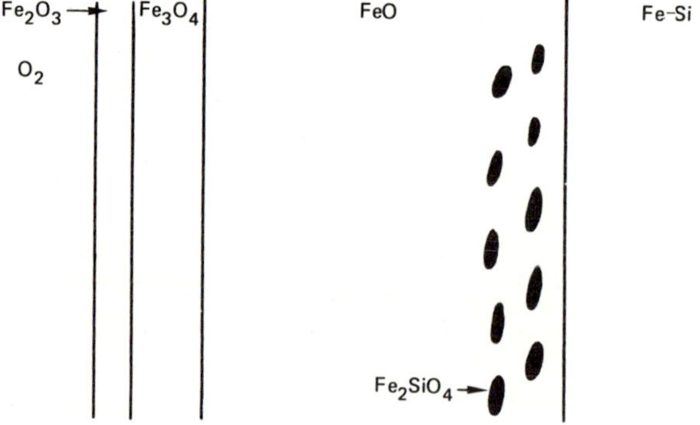

Fig. 5.20 Schematic diagram of the scale formed on dilute Fe–Si alloys

to segregate preferentially to the oxide formed. Where the alloy features internal oxidation as a reaction characteristic, the segregation is almost complete resulting in a matrix heavily deficient in the alloying element. Where an external scale only has formed although the segregation is severe, it does not generally lead to heavy denudation of the alloying component in the metal phase unless the alloy interdiffusion coefficient is very small. Furthermore, under conditions where internal oxidation would be expected a very rapid rate of external scale formation may reduce the thickness of the internal oxidation zone to negligible proportions and thus the denuded layer in the alloy does not develop.

A further classification in terms of behaviour and resulting morphology comes from the case when the alloying element is more noble than the parent element. A good example of this with commercial significance is given by the iron–copper system.

Fe–Cu alloys

When Fe–Cu alloys are oxidised, the copper does not enter the oxide phase but stays behind and enriches in the metal leading to the formation of a copper-rich rim at the scale–metal interface. This enrichment proceeds with continued scaling until the copper solubility limit in iron at that temperature is exceeded, when a second, metallic, copper-rich phase precipitates at the metal–scale interface. If the oxidation temperature is below the melting point of copper, this second phase is solid and, due to its low solubility for iron, may act as a barrier to iron diffusion resulting in slower rates of oxidation.

If, however, the scaling temperature is above the melting point of copper, the new phase is precipitated as a liquid. This liquid phase then penetrates inward along alloy grain boundaries. This situation can develop in the reheating of Cu-bearing steels before rolling. Copper may be present either as an impurity from the reduction of traces of copper minerals in iron ore in the blast furnace or remelting of Cu-bearing scrap or, intentionally, as an element to improve corrosion resistance. When such a steel is subsequently rolled, the liquid infiltrated grain boundaries, unable to support a shear or tensile stress, open up and the steel surface appears to be crazed. The resulting slab, billet, or ingot cannot usefully be processed and must be scrapped or returned for surface conditioning. This condition is known as 'hot shortness'. Hot shortness can also be encouraged by the presence in the steel of other relatively noble elements, such as Sn, As, Bi, and Sb, which also enrich with copper and lower the melting point of the second phase, thus aggravating the situation.

The problem of hot shortness can be approached in several ways. Initially, the copper level of the charge to a steelmaking furnace can be monitored and, if necessary, diluted with hot metal or pig iron. Secondly, the time–temperature history of the reheating programme can be scheduled to expose the material to sensitive temperatures for the shortest possible time. Thirdly, it is possible that heating at very high temperatures may result in back-diffusion of Cu being more rapid than scaling so the Cu concentration at the surface does not build up significantly. This is not a popular technique due to the excessive losses in yield due to scaling. Finally, the presence of nickel, which also enriches, serves to increase the solubility of copper in the enriched surface layers thus delaying precipitation. Alternatively, the phase precipitated may be nickel-rich and not liquid. Of these alternatives the first two, i.e. prevention at source and strictly controlled reheating procedures, are the most advantageous and, usually, the only economically feasible options.

Nb–Zr-type alloys

Alloys of Nb with small additions of Zr exhibit internal oxidation of Zr under an external scale of Nb-rich oxides. This class of alloy is somewhat different to those such as dilute Ni–Cr or Cu–Be alloys in that the external Nb-rich scale grows at a linear, rather than parabolic, rate. The kinetics of this process have been analysed by Rapp and Colson[27]. The analysis indicates the process should involve a diffusion controlled internal oxidation coupled with the linear scale growth, i.e. a paralinear process. At steady state, a limiting value for the penetration of the internal zone below the scale–metal interface is predicted. Rapp and Goldberg[28] have verified these predictions for Nb–Zr alloys.

Ni–Co alloys

Alloys of this type oxidise in a manner similar to pure nickel since CoO is only slightly more stable than NiO and the two oxides form a single phase solid solution scale[29]. However, the rate of oxidation is somewhat faster than that of pure nickel and a segregation of cations is observed across the scale. This segregation is illustrated in Fig. 5.21 which shows the concentration profiles for Co in the alloy and scale[29]. The increase in Co concentration near the scale–gas interface is the result of a higher mobility for Co ions than Ni ions through the oxide lattice. The higher mobility of Co is the result of two effects: a higher cation vacancy concentration in CoO than NiO, and a lower activation energy for the motion of the Co ions. The latter factor is consistent with calculations based on the difference in energy between a cation on an octahedral and tetrahedral site in a close packed oxygen anion lattice[30]. The most feasible diffusion path from one octahedral site to a vacant one is thought to be via an intervening tetrahedral site, which indicates why this energy difference affects the activation energy for cation motion. Since the Co-rich oxide has a higher cation vacancy concentration, as well as a lower activation energy for motion, the increase in CoO concentration near the scale–gas interface is responsible for the higher growth rate of the scale.

The partitioning of cations through a single phase scale due to different

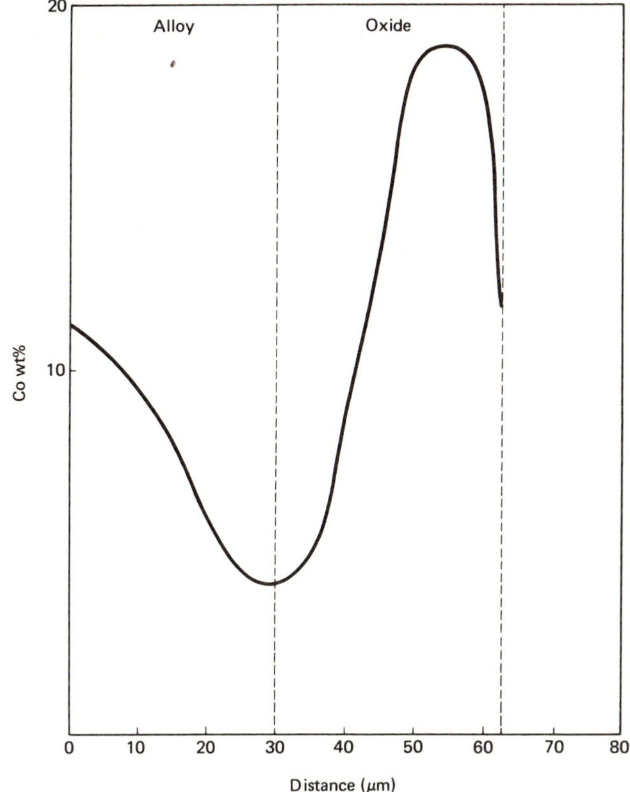

Fig. 5.21 Concentration profiles of Co across (Ni, Co)O scale formed on Ni–10.9Co alloy after 24 hours in 1 atm O_2 at 1000 °C

mobilities and the resultant effect of alloy composition on scale growth rate have been analysed by Wagner[31]. Bastow *et al.*[32] have measured concentration profiles and growth rates for a number of Ni–Co alloys and found close agreement with Wagner's model.

Additional factors in alloy oxidation

Stress development and relief in oxide films

The discussion of alloy oxidation in this chapter and the discussion of pure metal oxidation in the previous one have indicated that resistance to high temperature oxidation requires the development of an oxide barrier which separates the environment from the substrate. Continued resistance requires the maintenance of this protective barrier. Therefore, stress generation and relief in oxide films and the ability of an alloy to reform a protective scale, if stress-induced spalling or cracking occurs, are important considerations in the high temperature oxidation of metals. This subject has been discussed in reviews by Douglass[33], Stringer[34], and Hancock and Hurst[35].

Stress generation

The two principal sources of stress are *growth stresses*, which develop during the isothermal formation of the scale, and *thermal stresses*, which arise due to differential thermal expansion or contraction between the substrate and the scale.

Growth stresses

Growth stresses may occur due to a number of causes. The most important of these are

(a) volume differences between the oxide and the metal from which it forms,
(b) epitaxial stresses,
(c) compositional changes in the alloy or scale,
(d) point defect stresses,
(e) oxide formation within the scale,
(f) recrystallisation stresses,
(g) specimen geometry.

Each of these mechanisms will be discussed individually.

(*a*) *Volume differences between the oxide and the metal* The cause of stress in this case is due to the fact that the specific volume of the oxide is rarely the same as that of the metal which is consumed in its formation. The sign of the stress in the oxide may be related to the Pilling–Bedworth ratio[36].

$$\mathrm{PBR} = \frac{V\ \text{(per metal ion in oxide)}}{V\ \text{(per metal atom in metal)}} \tag{5.33}$$

Table 5.1 lists the PBRs for a number of systems[35]. The oxide is expected to be in compression if the PBR is greater than unity (which is the case for most metals) and in tension if the PBR is less than unity. Generally those systems which form tensile stresses in the oxide, e.g. K, Mg, Na, cannot maintain protective films. The oxides on most metals and alloys form in compression as expected from the PBR. However, this mechanism would only seem to be feasible if the oxide were growing at the scale–metal interface due to the inward migration of oxygen ions. Scales forming at the oxide–gas interface should not develop stresses due to the volume difference between metal and oxide. Also, the magnitudes of the stresses are not always in the order

Table 5.1 Oxide–metal volume ratios of some common metals[35]

Oxide	Oxide–metal volume ratio
K_2O	0.45
MgO	0.81
Na_2O	0.97
Al_2O_3	1.28
ThO_2	1.30
ZrO_2	1.56
Cu_2O	1.64
NiO	1.65
FeO (on α-Fe)	1.68
TiO_2	1.70–1.78
CoO	1.86
Cr_2O_3	2.07
Fe_3O_4 (on α-Fe)	2.10
Fe_2O_3 (on α-Fe)	2.14
Ta_2O_5	2.50
Nb_2O_5	2.68
V_2O_5	3.19
WoO_3	3.30

expected from the PBR[35]. Clearly, while the volume ratio may be a cause of growth stresses, other mechanisms must also be operative in many systems.

(*b*) *Epitaxial stresses* Nucleation considerations may dictate that the first oxide to form will have an epitaxial relationship with the substrate. This constraint will result in stress development due to the difference in lattice parameters between the metal and oxide. As the scale thickens, the epitaxial constraints are reduced so that this mechanism of stress generation is probably only significant for short oxidation times and/or low oxidation temperatures.

(*c*) *Compositional changes in the alloy or scale* Compositional changes can result in growth stresses by several means. Changes in the lattice parameter of the alloy as one or more elements are depleted by selective oxidation can produce stresses as can changes in the scale composition. Dissolution of oxygen in metals such as tantalum and niobium, which have high oxygen solubilities, can result in stress development. In a similar manner, the volume changes associated with internal oxide or carbide formation can result in stresses in some alloys.

(*d*) *Point defect stresses* Stresses can be generated in scales which exhibit large deviations from stoichiometry, e.g. FeO, due to the gradient in point defects across the scale which will result in lattice parameter variation across the scale. Also, metals which oxidise by the outward migration of cations may develop a vacancy gradient through the substrate due to injection from the scale. The lattice parameter variation produced by this gradient due to relaxation around the vacancies could result in stresses in the substrate. However, as pointed out by Hancock and Hurst[35], these vacancies may also be a source of stress relief by enhancing creep within the metal. However, the effects of this vacancy supersaturation are likely to be small since it is difficult to maintain and the vacancies tend to precipitate as voids at the scale–metal interface and grain boundaries in the substrate. This can be seen for the case of nickel in Fig. 5.22. The most important effect of vacancies appears to be that of decreasing the contact area between the scale and metal by void formation.

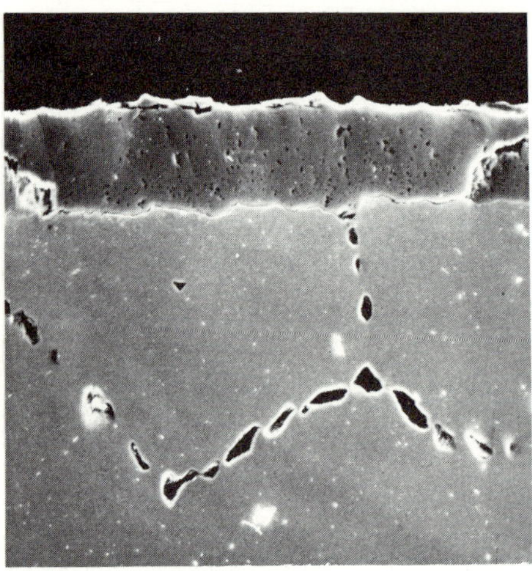

Fig. 5.22 Grain boundary voids formed in Ni due to vacancy condensation during oxidation at 950 °C

(*e*) *Oxide formation within the scale* The inward migration of oxidant along oxide grain boundaries and through microcracks can lead to compressive stress generation within the scale if it results in oxide formation at sites within the scale. This has been proposed by Rhines and Wolf[37] to explain the lengthening of Ni rods and the increase in area of Ni sheet during oxidation, both of which suggest compressive stresses in the scale. Also, Caplan and Sproule[38] found that areas of single crystal oxide formed on Cr were smooth, giving no indication of large stresses, whereas polycrystalline areas of oxide on the same specimens were buckled, indicating large compressive stresses within the scale. The possibility of new oxide forming within a scale is a matter of some controversy. However, it appears that this is a likely mechanism for compressive stress generation in scales on a number of metals and alloys.

(*f*) *Recrystallisation stresses* Recrystallisation in the oxide scale has been suggested as a cause of stresses[39,40]. However, this phenomenon would seem to relieve growth stresses rather than generate them. One similar case in which stress generation has been reported is for the oxidation of fine grained

Fe–Cr alloys[41]. Grain growth in the alloy was observed to disrupt the Cr_2O_3 scale locally and produce thick nodules of Fe-rich oxides. Oxidation of a coarse grained alloy produced a continuous Cr_2O_3 scale with no nodules.

(*g*) *Specimen geometry* The previously discussed mechanisms have applied for large, planar specimens. However, an important source of growth stresses arises due to the finite size of specimens and the resultant curvature. The nature of the stress developed will depend on the curvature and mechanism of scale growth. These effects have been divided into four categories by Hancock and Hurst[35]: cationic oxidation on convex surfaces, anionic oxidation on convex surfaces, cationic oxidation on concave surfaces, and anionic oxidation on concave surfaces. In the first case, when oxide grows on a convex surface by cation migration, the metal retreats as oxide forms at the scale–gas interface. If the scale remains adherent, compressive stresses are generated in the oxide as it tries to follow the metal surface. For the formation of scale on convex surfaces by anionic migration, the continued scale formation at the scale–metal interface can result in tensile stress development in the outer regions of the scale. When oxide forms on concave surfaces by cation migration, the initial oxide probably forms in compression but the geometry can result in a decrease and eventual change in sign of this stress so that, at longer times, the scale at the metal–oxide interface will be in tension. Finally, for oxide formation on concave surfaces by anion migration, the initial scale will form in compression and, as new scale forms at the scale–metal interface, these stresses will increase.

The above categorisation of specimen geometry effects is oversimplified but should serve to indicate the importance of this parameter in stress generation during isothermal oxidation.

Thermal stresses

Even when no stress exists at the oxidation temperature, stresses will be generated during cooling due to the difference in thermal expansion coefficients of the metal and oxide. The magnitude of this stress in the oxide may be expressed as[42]

$$\sigma_{ox} = \frac{E_{ox}\Delta T(\alpha_{ox} - \alpha_m)}{1 + 2\left(\frac{E_{ox}}{E_m}\frac{t_{ox}}{t_m}\right)} \tag{5.34}$$

where E = elastic modulus
α = coefficient of thermal expansion
t = thickness

The subscripts ox and m refer to the oxide and to the metal or alloy substrate, respectively. The coefficient of thermal expansion for the oxide will generally be less than that of the metal as seen in Table 5.2, so that compressive stresses

Table 5.2 Linear coefficients of thermal expansion of metals and oxides[35]

System	Oxide coefficient	Metal coefficient	Ratio
Fe/FeO	12.2×10^{-6}	15.3×10^{-6}	1.25
Fe/Fe_2O_3	14.9×10^{-6}	15.3×10^{-6}	1.03
Ni/NiO	17.1×10^{-6}	17.6×10^{-6}	1.03
Co/CoO	15.0×10^{-6}	14.0×10^{-6}	0.93
Cr/Cr_2O_3	7.3×10^{-6}	9.5×10^{-6}	1.30
Cu/Cu_2O	4.3×10^{-6}	18.6×10^{-6}	4.32
Cu/CuO	9.3×10^{-6}	18.6×10^{-6}	2.00

will develop in the oxide during cooling. The magnitude of the stress will be proportional to the difference in thermal expansion coefficients. This is consistent with observations that scales formed on metals such as nickel and cobalt tend to remain adherent after cooling, whereas those on Cu and Cr do not. Since the adhesive bond between metal and oxide is generally weaker than the cohesive bonds in either the metal or the scale, the result of large thermal stresses is generally spalling of oxide from the metal surface. This process is particularly detrimental to alloys which undergo severe depletion of the protective scale-forming elements during isothermal oxidation since subsequent exposures involve a less oxidation resistant alloy surface.

Stress relief

The growth and thermal stresses generated during oxidation may be accommodated by a number of mechanisms. The most important are

(a) cracking of the oxide,
(b) spalling of the oxide from the alloy substrates,
(c) plastic deformation of the substrate,
(d) plastic deformation of the oxide.

All of these mechanisms have been observed to operate in various systems. The particular mechanism will depend in a complicated manner on virtually all of the variables controlling the oxidation process.

Methods for improving scale adherence

There are three methods which have been used to improve scale adherence on metals and alloys. These are the addition of oxygen-active elements, e.g. Y, Hf, and the rare earths, to the alloys, the incorporation of an oxide dispersion, e.g. ThO_2, into the alloy, and the addition of noble metals, e.g. platinum. The third technique results in improved scale adherence[43] but requires rather large quantities of expensive metals so that it is limited to specific coatings applications. Furthermore, the origin of the effects are not well understood so this technique will not be discussed further.

The first two techniques have greater application and are somewhat related since the oxygen-active elements will be internally oxidised below the scale–alloy interface because of the high stability of their oxides. These techniques have been reviewed by Whittle and Stringer[44]. The effects of both oxygen-active metals and oxide dispersions occur for very small additions and include effects in addition to improved scale adherence, particularly for Cr_2O_3-forming alloys. Briefly these effects include (a) more rapid formation of a continuous Cr_2O_3 scale, (b) a change in the location of the scale-forming reaction from the scale–gas interface to the scale–metal interface, and (c) a reduction in the oxide grain size in many cases.

The effect of Ce on the scale adherence of a Ni–50Cr alloy is illustrated in the cyclic oxidation results of Fig. 5.23[45]. The alloy without Ce shows severe

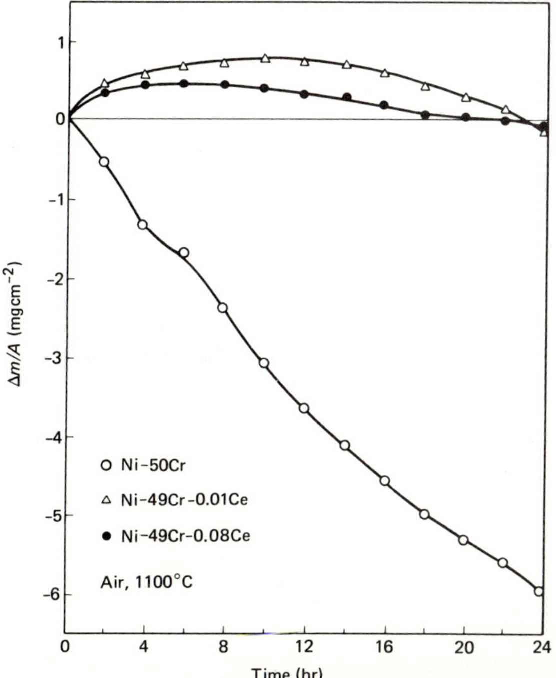

Fig. 5.23 Effect of cerium content on the cyclic oxidation mass loss of Ni–50Cr

mass losses due to scale spalling while the Ce-containing alloys show little or no indication of scale spalling, the long time mass losses being due at least partially to CrO_3 evaporation. The effect of Ce in decreasing the rate of isothermal scale growth on the same alloy is seen in Fig. 5.24 and the influence of Ce in maintaining a fine grained oxide is seen in Fig. 5.25.

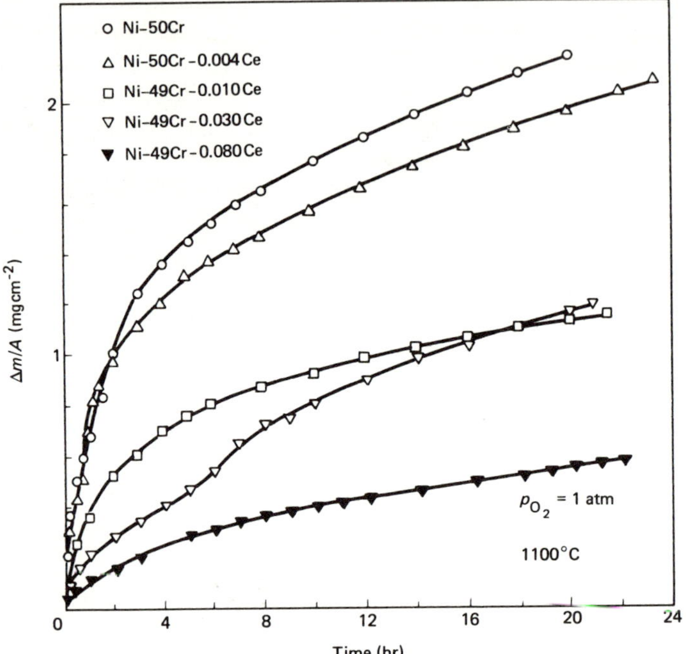

Fig. 5.24 Effect of cerium content on the oxidation rate of Ni–50Cr in oxygen

The major effect of additions to Al_2O_3-forming alloys appears to be improved scale adherence and minimisation of porosity at the scale–metal interface, although the transient oxidation period is also reduced. There appears to be little effect on the scale growth rate and, in fact, growth rate is sometimes slightly increased.

Numerous mechanisms have been proposed to explain the effects of oxygen-active elements and stable oxide dispersions. The most important effect on scale adherence appears to be the formation of pegs of the stable oxide which extend from the metal into the scale and key it to the surface. This is particularly true for Al_2O_3 scales, whereas a major additional factor for Cr_2O_3 scales is the decreased growth rate and accompanying lower stress levels. The most probable explanations for the effects on transient oxidation, oxide grain size, and scale growth rate for Cr_2O_3-forming alloys are associated with oxide nucleation and short-circuit transport paths. These effects were postulated by Stringer *et al.*[46] to be due to dispersoid particles acting as heterogeneous nucleation sites for oxide grains, reducing the internuclear distance, which allows more rapid formation of a continuous Cr_2O_3 film and produces a finer oxide grain size. The reduction in rate was explained as being due to the elimination of short-circuit diffusion paths for cations (probably

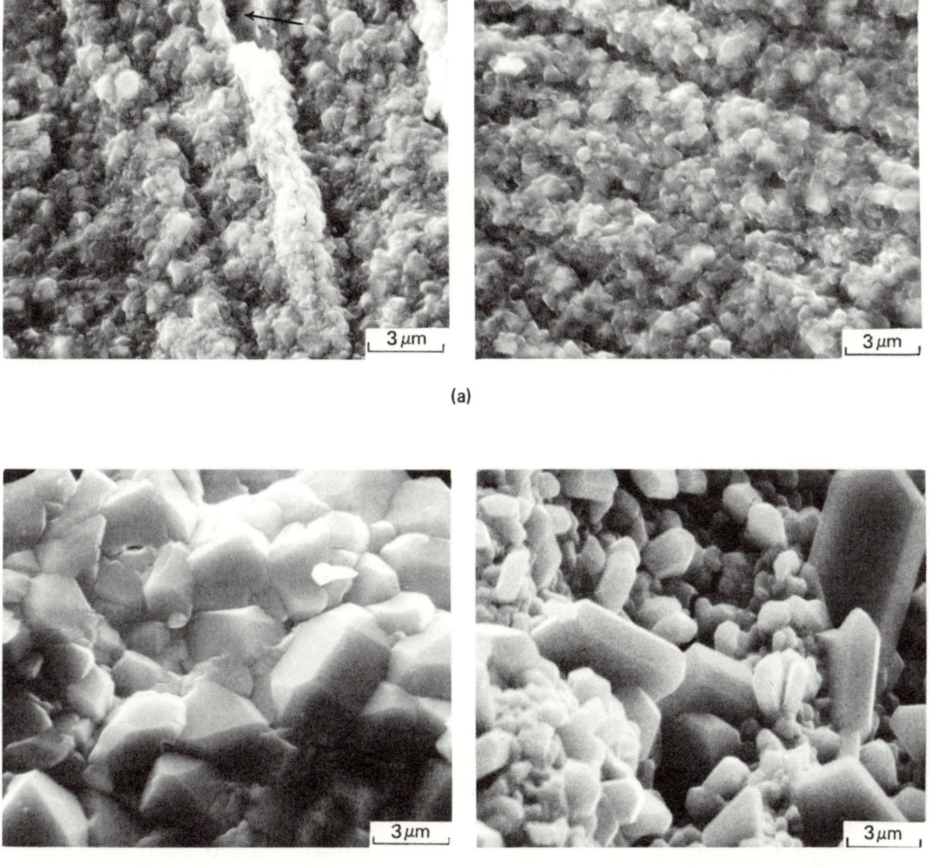

Fig. 5.25 Scanning electron micrograph showing the effect of cerium on slowing oxide grain growth. (a) after 1 minute at 1100 °C in oxygen the grain size of oxides grown on Ni–50Cr (left) and on Ni–49Cr–0.09Ce (right) is similar, but after 21 hours (b) the grain size of the oxide on Ni–50Cr (left) is much larger than that on Ni–49Cr–0.09Ce (right)

dislocations) so that anion diffusion becomes rate controlling. Ecer and Meier[45] later proposed that grain boundaries were the important short-circuit paths and that ions of the oxygen-active element addition blocked these paths.

The above discussion indicates that there is still considerable understanding to be gained regarding the mechanisms for the effects of oxygen-active elements. Also, their effects on scales other than Cr_2O_3 and Al_2O_3 are largely uninvestigated.

Catastrophic oxidation due to refractory metal additions

The oxidation literature contains several examples of catastrophic attack induced by high refractory metal contents in Fe- and Ni-base alloys. Leslie and Fontana[47] found that Mo-containing Fe–25Ni–16Cr alloys underwent catastrophic oxidation at 900 °C in static air even though the oxidation resistance was good in flowing air. Brenner[48] found that binary Ni–Mo and Fe–Mo alloys containing as much as 30% Mo did not oxidise catastrophically at 1000 °C. However, a number of Fe–Ni–Mo and Fe–Cr–Mo alloys did undergo catastrophic attack[49]. The mechanism postulated for this attack consists of the formation of a layer of MoO_2 at the scale–alloy interface, which is oxidised to liquid MoO_3 following cracking of the scale. (MoO_3 melts at 801 °C but forms low melting eutectics with most oxides.) The molten oxide may result in the dissolution and disruption of the protective scale.

The postulate of a liquid phase being involved in the catastrophic degradation is consistent with the earlier work of Rathenau and Meijering[50], which indicated that Mo-induced accelerated attack commenced at temperatures approximating the eutectic temperatures for the oxides formed on the alloys and Mo oxides. The corrosion of Cr_2O_3-forming alloys was associated with Cr_2O_3 dissolution by liquid Mo oxides.

A further important feature of Mo-induced attack is the high volatility of the Mo oxides as was illustrated in Fig. 4.13 for a temperature of 1250 K. The influence of this was observed in the work of Peters *et al.*[51] who found that Ni–15Cr–Mo alloys with Mo-contents in excess of 3% were catastrophically attacked in static oxygen at 900 °C. A Mo-rich oxide was observed at the scale–alloy interface suggesting the importance of MoO_3 accumulation. The same alloy, however, formed a protective scale in rapidly flowing oxygen and oxidised at about the same rate as a Ni–15Cr alloy without Mo. Apparently, the evaporation of MoO_3 in the flowing atmosphere prevented sufficient accumulation of Mo oxides to cause severe corrosion.

Tungsten additions have not been observed to cause catastrophic oxidation of Ni- and Co-base alloys, perhaps due to the higher melting temperatures of the W oxides as compared with Mo oxides, e.g. T_{mp} = 1745 K for WO_3. However, W additions have been observed to induce some scale breakdown on Ni–Cr alloys[52]. Additions of W to Co–Cr alloys appear to be beneficial in decreasing the transient oxidation period and promoting the formation of a continuous Cr_2O_3 layer[53].

Relatively little work has been done on the effects of refractory metal additions on the oxidation of Al_2O_3-forming alloys although they have been observed to initiate severe corrosion when a molten sulphate deposit is on the alloy[54]. This phenomenon will be discussed in Chapter 7.

Coatings for oxidation protection

The discussion of this and the previous chapter have made it clear that many of the materials required for high temperature applications do not have adequate oxidation resistance. The refractory metals, such as Mo and W, oxidise rapidly due to the formation of volatile oxides, Ta and Nb have a large oxygen solubility and scales which continually crack, plain carbon steels form iron-rich oxides which grow at an unacceptable rate, and most Ni- and Co-base alloys have mechanical property-dictated limitations on their compositions which preclude advantage being taken of the optimum oxidation resistance capable of development in these systems. These problems are generally approached by the use of coatings which do not have, or require, the same mechanical properties as the substrate but provide oxidation resistance by the selective oxidation of elements, e.g. Al, Cr, or Si.

The detailed requirements for a coating depend on the particular application for which it is intended. However, these requirements can be catalogued in a general way as follows. The coating must contain elements which will provide isothermal oxidation resistance by selective oxidation in the environment. In addition, the fact that most applications involve temperature changes requires that the coating be resistant to cyclic oxidation as well. Variations in temperature also require that the coating be resistant to thermal fatigue and that its coefficient of thermal expansion be sufficiently close to that of the substrate to avoid spalling of the coating. In some applications the coating must also be resistant to erosion by particulates. Finally, the coating should have no adverse effects on the room-temperature or elevated-temperature mechanical properties of the substrate.

Coatings may be applied by a number of techniques which have been reviewed by Chatterji *et al.*[55]. These include mechanical cladding, hot dipping, hot metal spraying, slurry coating, electrophoresis, electroplating, metalliding, vapour deposition, plasma spraying, sputtering, glow discharge impregnation, and pack cementation.

There are two basic types of coating applied for oxidation resistance. These are diffusion coatings, in which one or more elements are diffused into the substrate surface to form an oxidation resistant layer, and overlay coatings, in which a layer of an oxidation resistant alloy is applied to the substrate with a minimum of interdiffusion. The best examples of diffusion coatings are silicide coatings on the refractory metals Mo and W[55] and aluminide coatings on Ni- and Co-base superalloys[56]. In the former case, Si is deposited on the surface and then diffused in to form molybdenum and tungsten silicides, e.g. $MoSi_2$ and WSi_2. These compounds oxidise to form Si-rich oxides which provide much improved oxidation resistance. In the case of aluminide coatings, Al is diffused into the surfaces of superalloys to produce NiAl and CoAl. These compounds then form protective Al_2O_3 scales during oxidation.

The diffusion coatings involve the substrate alloy in the formation of the coating and are, therefore, somewhat limited in the compositions which can be obtained. The second type of coating, the overlay, circumvents this problem to a large extent. The most important overlay coatings are the MCrAlY materials where M is one or more of the elements Fe, Ni, or Co. The yttrium is present in small concentrations to improve scale adherence. These coatings form Al_2O_3 scales by selective oxidation and comprise the most advanced protective systems for superalloys at the present time.

The oxidation behaviour of coatings follows the same principles of alloy oxidation discussed earlier in this chapter, i.e. they protect by selective oxidation to form Al_2O_3, Cr_2O_3, or SiO_2 scales. The oxidation behaviour of coatings is rather more complex, however, because of their thinness. This means that depletion of the element being selectively oxidised is an important phenomenon, particularly when scale spalling occurs during thermal cycling. An additional factor, which can lead to enhanced degradation, is interdiffusion between the coating and the substrate. This can lead to depletion of important elements, such as Si and Al, and to detrimental elements entering the coating from the substrate.

References

1 Kubaschewski, O. and Hopkins, B. E., *Oxidation of Metals and Alloys*, Butterworth, London, 1962

2 Hauffe, K., *Oxydation von Metallen und Metallegierungen*, Springer, Berlin, 1957

3 Bénard, J., *Oxydation des Métaux*, Gauthier-Villars, Paris, 1962

4 Pfeiffer, H. and Thomas, H., *Zunderfeste Legierungen*, Springer, Berlin, 1963

5 Kofstad, P., *High Temperature Oxidation of Metals*, Wiley, New York, 1966

6 Birchenall, C. E., Oxidation of alloys, in: *Oxidation of Metals and Alloys*, ed. D. L. Douglass, ASM, Metals Park, Ohio, 1971; Chapter 10

7 Mrowec, S. and Werber, T., *Gas Corrosion of Metals*, National Bureau of Standards and National Science Foundation, translated from Polish, 1978

8 Wagner, C., *Ber. Bunsenges. phys. Chem.*, **63,** 772, 1959

9 Wagner, C., *J. Electrochem. Soc.*, **99,** 369, 1952

10 Wagner, C., *J. Electrochem. Soc.*, **103,** 571, 1956

11 Rickert, H., *Z. phys. Chem. N. F.*, **21,** 432, 1960

12 Rapp, R. A., *Corrosion*, **21,** 382, 1965

13 Swisher, J. H., Internal oxidation, in: *Oxidation of Metals and Alloys*, ed. D. L. Douglass, ASM, Metals Park, Ohio, 1971; Chapter 12

14 Meijering, J. L., Internal oxidation in alloys, in: *Advances in Materials Research*, Volume 5, ed. H. Herman, Wiley, New York, 1971; p. 1–81
15 Wood, S., Adamonis, D., Guha, A., Soffa, W. A. and Meier, G. H., *Met Trans.*, **6A,** 1793, 1975
16 Megusar, J. and Meier, G. H., *Met. Trans.*, **7A,** 1133, 1976
17 Bohm, G. and Kahlweit, M., *Acta Met.*, **12,** 641, 1964
18 Bolsaitis, P. and Kahlweit, M., *Acta Met.*, **15,** 765, 1967
19 Wagner, C., *Z. Elektrochem.*, **63,** 772, 1959
20 Rapp, R. A., *Acta Met.*, **9,** 730, 1961
21 Wagner, C., *Corros. Sci.*, **5,** 751, 1965
22 Pickering, H. R., *J. Electrochem. Soc.*, **119,** 64, 1972
23 Maak, F., *Z. Metallkunde*, **52,** 545, 1961
24 Giggins, C. S. and Pettit, F. S., *TAIME*, **245,** 2495, 1969
25 Birks, N. and Rickert, H., *J. Inst. Met.*, **91,** 308, 1962–63
26 Ecer, G. M. and Meier, G. H., *Oxid. Metals*, **13,** 119, 1979
27 Rapp, R. A. and Colson, H., *TAIME*, **236,** 1616, 1966
28 Rapp, R. A. and Goldberg, G., *TAIME*, **236,** 1619, 1966
29 Wood, G. C., The structures of thick scales on alloys, in: *Oxidation of Metals and Alloys*, ed. D. L. Douglass, ASM, Metals Park, Ohio, 1971; Chapter 11
30 Cox, M. G. C., McEnaney, B. and Scott, V. D., *Phil. Mag.*, **26,** 839, 1972
31 Wagner, C., *Corros. Sci.*, **9,** 91, 1969
32 Bastow, B. D., Whittle, D. P. and Wood, G. C., *Corros. Sci.*, **16,** 57, 1976
33 Douglass, D. L., Exfoliation and the mechanical behavior of scales, in: *Oxidation of Metals and Alloys*, ed. D. L. Douglass, ASM, Metals Park, Ohio, 1971
34 Stringer, J., *Corros. Sci.*, **10,** 513, 1970
35 Hancock, P. and Hurst, R. C., The mechanical properties and breakdown of surface oxide films at elevated temperatures, in: *Advances in Corrosion Science and Technology*, ed. R. W. Staehle and M. G. Fontana, Plenum Press, New York, 1974
36 Pilling, N. B. and Bedworth, R. E., *J. Inst. Met.*, **29,** 529, 1923
37 Rhines, F. N. and Wolf, J. S., *Met. Trans.*, **1,** 1701, 1970
38 Caplan, D. and Sproule, G. I., *Oxid. Metals*, **9,** 459, 1975
39 Jaenicke, W. and Leistikow, S., *Z. phys. Chem.*, **15,** 175, 1958
40 Jaenicke, W., Leistikow, S. and Stadler, A., *J. Electrochem. Soc.*, **111,** 1031, 1964
41 Horibe, S. and Nakayama, T., *Corros. Sci.*, **15,** 589, 1975
42 Oxx, G. D., *Prod. Eng.*, **29,** 61, 1958
43 Felten, E. J. and Pettit, F. S., *Oxid. Metals*, **10,** 189, 1976
44 Whittle, D. P. and Stringer, J., *Phil. Trans. Roy. Soc. London*, **295A,** 309, 1980

45 Ecer, G. M. and Meier, G. H., *Oxid. Metals*, **13,** 159, 1979
46 Stringer, J., Wilcox, B. A. and Jaffee, R. I., *Oxid. Metals*, **5,** 11, 1972
47 Leslie, W. C. and Fontana, M. G., *Trans. ASM*, **41,** 1213, 1949
48 Brenner, S. S., *J. Electrochem. Soc.*, **102,** 7, 1955
49 Brenner, S. S., *J. Electrochem. Soc.*, **102,** 16, 1955
50 Rathenau, G. W. and Meijering, J. L., *Metallurgia*, **42,** 167, 1950
51 Peters, K. R., Whittle, D. P. and Stringer, J., *Corros. Sci.*, **16,** 791, 1976
52 El-Dashan, M. E., Whittle, D. P. and Stringer, J., *Corros. Sci.*, **16,** 83, 1976
53 El-Dashan, M. E., Whittle, D. P. and Stringer, J., *Corros. Sci.*, **16,** 77, 1976
54 Goebel, J. A., Pettit, F. S. and Goward, G. W., *Met. Trans.*, **4,** 261, 1973
55 Chatterji, D., DeVries, R. C. and Romeo, G., Protection of superalloys for turbine application, in: *Advances in Corrosion Science and Technology*, ed. M. G. Fontana and R. W. Staehle, Plenum Press, New York, 1976; Chapter 1
56 Goward, G. W., Protective coatings for high temperature alloys – state of technology, in: *Properties of High Temperature Alloys*, ed. Z. A. Foroulis and F. S. Pettit, Electrochem. Soc., Princeton, New Jersey, 1976; p. 806

6
Reaction of metals in mixed environments

Introduction

This chapter deals with the reactions and reaction mechanisms that arise when a metal is exposed to an atmosphere containing more than one species with which the metal can form a compound. This condition is, of course, found in most commercial atmospheres.

The problem can be considered under three main headings.

(a) A complex scale is formed consisting of two or more reaction products which form directly by reaction with the atmosphere.
(b) Initial reaction leads to formation of a scale of one compound, usually the oxide, with compounds of the second reactant being found at the metal–scale interface or within the metal. Alternatively, the second reactant simply dissolves in the metal. In this case, the mechanism involves penetration of the first formed scale by the second species.
(c) The outward diffusion of solute cations through the scales to form compounds with the second reactant at the scale–gas interface.

Usually these situations arise when the metal is exposed to an atmosphere that is primarily oxidising (i.e. oxide forming) but contains other elements with which the metal can react. Examples are air, when the metal shows solubility for nitrogen or can form stable nitrides, and atmospheres containing water vapour or carbon dioxide (common in combustion atmospheres) which can lead to hydrogen and carbon dissolution, carbide formation, and even carbon deposition. Particularly severe reaction can result when sulphur is present in the atmosphere and, since most metals will form sulphides, this results in rapid reactions due to the formation of liquid phases, or to the removal of the protective elements from an alloy in the form of discrete sulphide particles, or simply due to rapid cation diffusion through the sulphide phase.

Iron–sulphur–oxygen system

Complete understanding of the processes involves both thermodynamic and kinetic aspects of the mechanisms, and the subject is conveniently approached by consideration of the reaction of a metal with an atmosphere containing both sulphur and oxygen[1–3]. To avoid unnecessary confusion it is also convenient to consider a particular case, such as the iron–sulphur–oxygen system, since the principles involved can be applied in identical manner to other systems. This avoids the confusion of anonymous symbols and fictitious compounds.

The first point to consider is whether both oxide and sulphide can form together, but independently, from the atmosphere. This means that three conditions must be met.

(a) The oxygen partial pressure of the atmosphere must be higher than the decomposition oxygen pressure of the oxide.
(b) The sulphur partial pressure must be higher than the decomposition sulphur partial pressure of the sulphide.
(c) Oxide and sulphide must coexist in equilibrium with the atmosphere.

Consider the iron–oxygen–sulphur system; the first two conditions are easy to meet and can be expressed as

$$\text{(a)}\quad Fe + \tfrac{1}{2}O_2 = FeO;\ \Delta G_1^\ominus = -259\,370 + 62.5T\,\text{J} \qquad (6.1)$$

$$\therefore\quad (p_{O_2})\ \text{atmosphere} > \frac{1}{a_{Fe}^2}\exp\left(2\Delta G_1^\ominus/RT\right)$$

$$\text{i.e.}\qquad > \frac{1}{a_{Fe}^2}\exp\left(15.32 - 63\,590/T\right)$$

$$\text{(b)}\quad Fe + \tfrac{1}{2}S_2 = FeS;\ \Delta G_2^\ominus = -150\,100 + 51.5T\,\text{J} \qquad (6.2)$$

$$\therefore\quad (p_{S_2})\ \text{atmosphere} > \frac{1}{a_{Fe}^2}\exp\left(2\Delta G_2^\ominus/RT\right)$$

$$\text{i.e.}\qquad > \frac{1}{a_{Fe}^2}\exp\left(12.63 - 36\,800/T\right)$$

The third condition can be expressed as

$$\text{(c)}\quad FeO + \tfrac{1}{2}S_2 = FeS + \tfrac{1}{2}O_2;\ \Delta G_3^\ominus = 109\,270 - 11.0T\ \text{J} \qquad (6.3)$$

and for equilibrium between FeO and FeS the composition of the atmosphere is constrained to certain compositions expressed by

$$(p_{O_2}/p_{S_2})_{FeO/FeS} = \exp\left(2.40–26\,272/T\right)$$

This condition is not easy to meet. The situation is best understood by consideration of the Fe–S–O phase diagram given in Fig. 6.1, where the conditions (a), (b), and (c) correspond to atmospheres as follows.

(a) corresponds to atmospheres within the FeO and Fe_3O_4 phase fields, i.e. to the right of line ACDE,

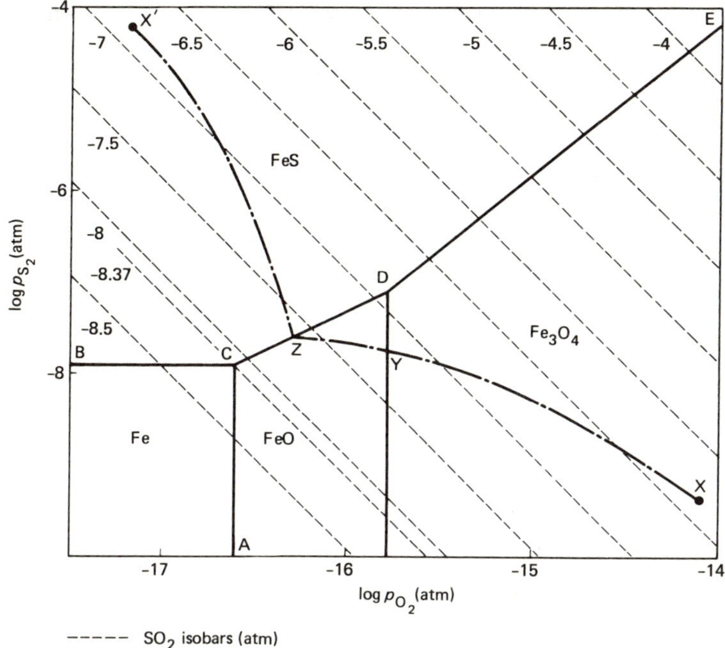

Fig. 6.1 The iron–sulphur–oxygen phase diagram at 900 °C showing reaction paths for duplex scale formation

(b) corresponds to atmospheres within the FeS phase field, i.e. above line BCDE,

(c) corresponds only to atmospheres whose compositions lie on the line CDE.

Since condition (c) could only be met by careful and accurate control of the atmosphere composition, it is unlikely that a mixed (or duplex) scale would be formed while the scale–gas interface maintains equilibrium with the bulk atmosphere composition.

From Fig. 6.1 it can be seen that many atmospheres correspond to conditions under which either oxide or sulphide is the stable phase. Consider a piece of iron held in an atmosphere within the Fe_3O_4 field at position X. Initially Fe_3O_4 will form, however the reaction will denude the atmosphere of oxygen in the region of the specimen surface resulting in a local change of atmosphere there to lower oxygen partial pressures. In atmospheres containing sulphur dioxide, this results in a local increase in sulphur partial pressure since the sulphur, oxygen, and sulphur dioxide partial pressures are controlled by the equilibrium

$$\tfrac{1}{2}S_2 + O_2 = SO_2;\ \Delta G_4^{\ominus} = -364\,000 + 72.6T\ \text{J} \tag{6.4}$$

Thus, the gas composition at the scale–gas interface (or the metal surface in these early stages) follows the path XYZ, entering the FeO phase field at point Y. At this stage the Fe_3O_4, formed earlier, is reduced to FeO and further reaction proceeds according to

$$2Fe + SO_2 = 2FeO + \tfrac{1}{2}S_2; \ \Delta G_5^\ominus = 2\Delta G_1^\ominus - \Delta G_4^\ominus \tag{6.5}$$

consuming SO_2, forming FeO, and releasing S_2 into the boundary layer at the specimen surface where the composition thereby changes along the path YZ. On reaching Z both oxide, FeO, and sulphide, FeS, may form together. Since SO_2 is being consumed, the reaction path XYZ must cross the SO_2 isobars which are shown as broken lines. This represents the establishment of a sulphur dioxide partial pressure gradient across the boundary layer in the gas at the specimen surface, as shown in Fig. 6.2. SO_2 is supplied to the specimen by diffusing from the bulk gas through the boundary layer.

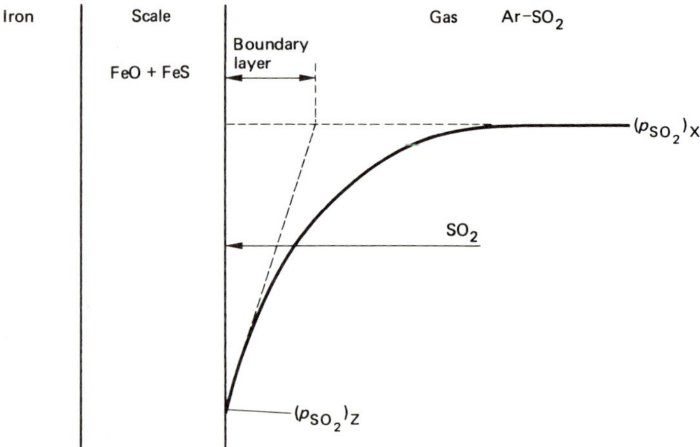

Fig. 6.2 Formation of a duplex oxide–sulphide scale on iron in an Ar–SO_2 atmosphere. Rapid reaction depletes the SO_2 at the scale–gas interface establishing a gradient of SO_2 in the boundary layer

Thermodynamically, this means that the local sulphur partial pressure at the scale surface is given by the equilibrium of reaction 6.5

$$\frac{p_{S_2}^{*1/2}\, a_{FeO}^{*2}}{a_{Fe}^{*2}\, p_{SO_2}^{*}} = \exp\left(-\Delta G_5^\ominus/RT\right) \tag{6.6}$$

i.e.

$$p_{S_2}^{*} = a_{Fe}^{*4}\, p_{SO_2}^{*2} \exp\left(-2\Delta G_5^\ominus/RT\right) \tag{6.7}$$

assuming that $a_{FeO}^{*} = 1$. The asterisk (*) denotes values at the scale–gas interface.

According to equation 6.2, FeS can form if

$$p^*_{S_2} > \frac{1}{a^{*2}_{Fe}} \exp\,(2\Delta G^{\ominus}_2/RT)$$

i.e. if

$$a^{*4}_{Fe}\, p^{*2}_{SO_2} \exp\,(-\,2\Delta G^{\ominus}_5/RT) > \frac{1}{a^{*2}_{Fe}} \exp\,(2\Delta G^{\ominus}_2/RT)$$

Rearranging and substituting $\Delta G^{\ominus}_5 = 2\Delta G^{\ominus}_1 - \Delta G^{\ominus}_4$, this becomes

$$a^{*3}_{Fe}\, p^*_{SO_2} \exp\,[(\Delta G^{\ominus}_4 - \Delta G^{\ominus}_2 - 2\Delta G^{\ominus}_1)/RT] > 1 \qquad (6.8)$$

Thus, the formation of a duplex scale depends upon modification of the local atmosphere around the specimen to provide conditions which will support the formation of both compounds. This can occur only when the rate controlling process is diffusion through the gas phase or when a phase boundary reaction is the rate controlling step. In either case, a constant reaction rate would be observed and, as the duplex scale thickens, the value a^*_{Fe} at the scale–gas interface must fall, eventually reaching such a low value that equation 6.8 is no longer satisfied, duplex scale formation ceases, and an outer layer of FeO forms as diffusion of cations through the scale becomes the rate determining step for the reaction.

Equally evident from equation 6.8 is that, below a critical value of sulphur dioxide partial pressure, sulphide formation cannot occur. This can be calculated by putting $a^*_{Fe} = 1$ in equation 6.8 and corresponds to the SO_2 isobar passing through point C in Fig. 6.1. Precisely the same argument may be used for an atmosphere whose bulk composition lies in the FeS field when the sulphide forms first using reaction path X′Y′Z in Fig. 6.1.

The morphology of the duplex scale can vary from a lamellar structure of alternate FeS and FeO lamellae to a seemingly random mixture of sulphide in oxide[4]. It is likely, however, that in all cases the sulphide phase is continuous otherwise it could not maintain the high transport rates to support the rapid reactions observed in these cases.

The detailed mechanism of the reaction is given in Fig. 6.3. The lamellar structure is used to represent the continuous sulphide network and it can be seen that the sulphide provides a means of rapid cation transport which maintains a high value of metal activity at the surface of the scale (a^*_{Fe}) by cross-feeding cations into the neighbouring oxide lamellae. High metal activity is maintained there also. At the scale surface, oxide forms by reaction with the SO_2 molecules releasing sulphur which then diffuses, by surface diffusion or via the gas phase, to form FeS on the appropriate lamellae. Thus the mechanism is a cooperative one in which the necessary cation supply is provided by the sulphide phase whereas the conditions conductive to sulphide formation are maintained by the formation of the oxide phase.

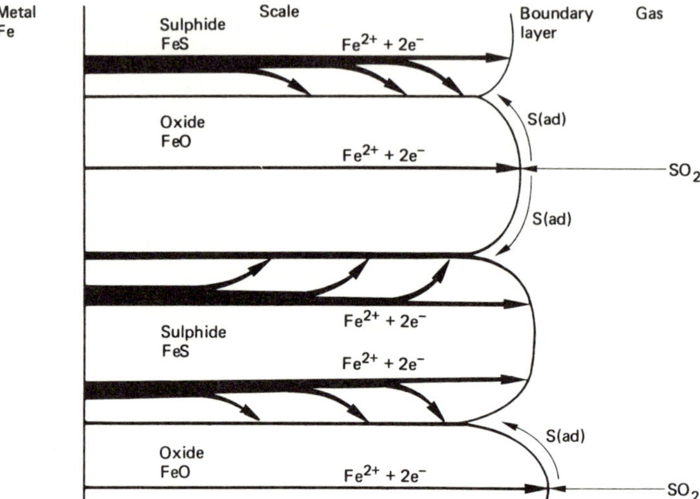

Fig. 6.3 Mechanism of formation of duplex scale (FeO + FeS) on iron from Ar–SO_2 atmospheres, showing how the FeS lamellae provide rapid cation transport and also supply the adjacent FeO lamellae

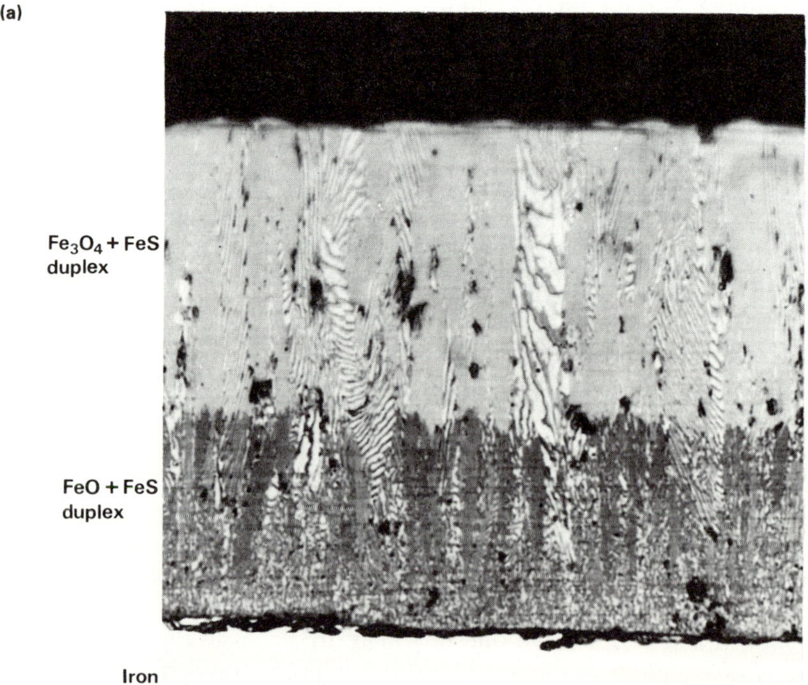

Figure 6.4 shows examples of duplex scale formed by this mechanism on iron[4], nickel[5,7], and cobalt[6] in Ar–SO_2 atmospheres, i.e. following path XYZ in Fig. 6.1. The lamellar structure on iron has also been found on iron exposed to C–O–S atmospheres[3], following path X′Z, confirming the general nature of the above analysis and mechanisms[2].

The presence of SO_2 in commercial atmospheres has long been known to result in increased reaction rates and the formation of sulphides close to the metal surface. Recently it has also been found that the attack of alloys in CO_2

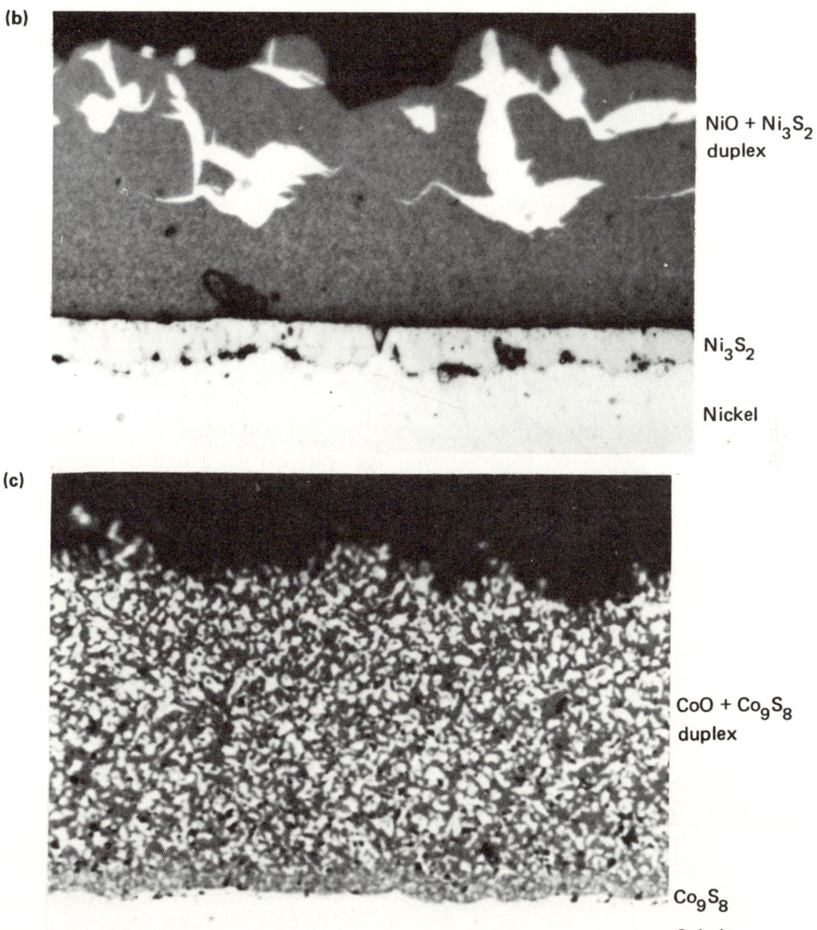

Fig. 6.4 Examples of duplex scales: (a) duplex scale formed on iron heated to 900 °C in Ar–1% SO_2. ×800, (b) massive duplex scale formed on nickel heated to 500 °C in Ar–1% SO_2. ×600, (c) fine duplex scale formed on cobalt held at 840 °C in Ar–2% SO_2. ×500

atmospheres has led to carbide formation and scale bursting due to carbon deposition within the scale. This is initially surprising because, in such cases, a protective oxide scale had formed and, according to the above analysis, sulphide (or carbide) formation certainly could not occur at the scale–gas interface, where the metal activity is low, in equilibrium with the gas phase. However, the metal activity increases from the scale–gas interface to the scale–metal interface and, if the sulphur or carbon species can penetrate the scale, it will eventually reach a part of the system where the metal activity is high enough to form sulphide or carbide[8]. The metal–scale interface has the highest metal activity and is, therefore, the site most amenable to sulphide or carbide formation.

There are two ways in which sulphur or carbon may be transported through a protective scale. It may dissolve and diffuse through the oxide lattice (solution–diffusion) or it may transport as a gas down pores and microcracks (gas permeation)[8].

It is not possible to eliminate the solution–diffusion mechanism although data on the solubility of sulphur in oxides (~0.01 wt%) indicate that the process may be slow. More important is that, thermodynamically, the mechanism can only occur under a restricted range of conditions where the sulphur partial pressure in the atmosphere is greater than that in equilibrium with metal and metal sulphide. This condition is given in equation 6.2 and corresponds to atmospheres in the FeO phase field above the extrapolation of line BC in Fig. 6.1.

Alternatively, the permeation of, say, SO_2 molecules through the protective oxide scale may occur so long as the SO_2 partial pressure in the gas is greater than the limiting value obtained by rearranging equation 6.8 and putting $a^*_{Fe} = 1$

$$p_{SO_2} > \exp\left[(\Delta G_2^{\ominus} + 2\Delta G_1^{\ominus} - \Delta G_4^{\ominus})/RT\right] \tag{6.9}$$

This of course corresponds to the SO_2 isobar which passes through point C in Fig. 6.1. Thus sulphide formation at the scale–metal interface is possible in atmospheres which are highly oxidising but which contain a small amount of SO_2. The limiting amount of SO_2 varies from about 4% for the Ni–S–O system down to extremely low values for metals such as iron (10^{-7} atm) and chromium (10^{-16} atm). This is easily calculated by using the data appropriate to the system in equation 6.9. The validity of this mechanism has been demonstrated in the cases of nickel and cobalt exposed to sulphurous atmospheres after preoxidation in sulphur-free atmospheres[9,10].

Other environments

The analysis of carbon penetration of oxide layers with subsequent carbon dissolution, carbide precipitation in appropriate alloys, and carbon deposition within the scale is analogous to the above. The carbonaceous gas, i.e. CO_2, enters the scale and migrates down microcracks towards the metal–scale interface. On the way it modifies its composition to conform to the oxygen potential at that position within the scale according to the equilibrium

$$CO_2 = CO + \tfrac{1}{2}O_2; \ \Delta G^{\ominus}_{10} = 282\,150 - 86.7T \text{ J} \qquad (6.10)$$

Since the oxygen partial pressure falls continuously across the scale to a value in equilibrium with the metal and oxide at the metal–scale interface, the CO/CO_2 ratio will similarly change, becoming richer in CO as it approaches the scale–metal interface. At the scale–metal interface, if this is a site of porous oxide formation, the CO/CO_2 redox mixture may enter into a series of reactions ferrying oxygen across the pore as below.

at the oxide surface $\quad CO + MO = CO_2 + M \qquad$ (6.11)
at the metal surface $\quad CO_2 + M = CO + MO \qquad$ (6.12)

The metal of equation 6.11 is released as ions and electrons which migrate through the scale towards the metal–scale interface. This is shown in Fig. 6.5.

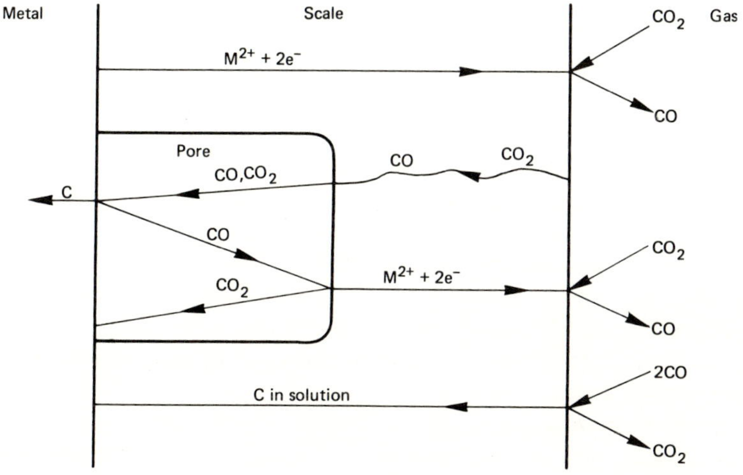

Fig. 6.5 Reactions involving CO and CO_2, including transport through the scale and oxygen ferrying across pores

It is also possible that carbon may dissolve in the metal during reaction there and the reaction at the metal surface may be represented as

$$CO_2 + 2M = \underline{C} + 2MO \qquad (6.13)$$

If the metal is an alloy which contains elements that form stable carbides, then it is possible that metal carbides may precipitate in the surface layers of the metal as the concentration of dissolved carbon increases and exceeds the solubility limit of the carbide. The more stable the carbide, and the higher the concentration of the metal in solution, the lower is the carbon concentration required to initiate precipitation.

This effect is particularly harmful if the metal forming the carbide within the alloy is also the metal whose oxide forms the protective scale. For example, this occurs with alloys which depend on chromium for their protection. The CO/CO_2 system penetrates the protective Cr_2O_3 scale and carbon dissolves in the alloy eventually forming discrete chromium carbide precipitates. This removes the chromium from general solution and redistributes it as chromium carbides. When the carbides eventually oxidise it is no longer possible to form a complete layer of Cr_2O_3 and a non-protective scale results. For this reason, chromium steels are susceptible to high oxidation rates when exposed to carbonaceous atmospheres for long periods at high temperature.

Finally, at lower temperatures (500 °C), the enrichment of the gas in the pore space within the scale in carbon monoxide, as it penetrates towards the metal–scale interface, leads to deposition of carbon within the scale by the Bouduard reaction

$$2CO = CO_2 + C; \Delta G_{14}^{\ominus} = -170\,290 + 174.26T \text{ J} \qquad (6.14)$$

The deposition of carbon leads to swelling and, eventually, bursting of the protective scale which thus loses its protective nature and rapid reactions ensue.

Fortunately, the underlying metal can act as a sink for the carbon[11] especially if it contains carbide-forming alloying elements. The entrapment of carbon in the metal thus prevents the build up of carbon monoxide within the scale cavities and delays carbon deposition. Thus, so long as the capacity of the metal for carbon is high enough to continue absorbing the carbon over its designed life, bursting of the scale by this mechanism should not be a problem. It is, however, readily observed in the laboratory with thin section metal specimens of low carbon capacity.

A further related phenomenon is that of 'metal dusting' which occurs in atmospheres of high carbon content and low oxygen potential, i.e. atmospheres typical of reforming processes. The carbide formed within the metal decomposes to form graphite; the associated expansion causing both metal and carbon to flake away from the surface. In spite of the high carbon activity at the metal surface, carbon is not seen to deposit there, which is due, presumably, to nucleation difficulties combined with rapid dissolution into the metal. This phenomenon is observed with iron, nickel, and cobalt and their alloys. The formation of compact oxide scales does not prevent this reaction but only causes an incubation time to be observed.

In the case of alloys containing alloying elements which form stable carbides, the incoming carbon reacts to precipitate these carbides and then forms metastable carbides of the matrix metal which then decomposes to graphite causing metal dusting as described above. This is found in Ni–Cr and Fe–Ni–Cr alloys in CO-rich atmospheres typical of reforming and gas synthesis operations. The eventual appearance of the light green porous nickel oxide in attacked regions led to the name 'green rot' being applied. This is due to the removal of chromium from general solution, breakdown of a Cr_2O_3 protective layer, and formation of nickel oxide at such suitable sites.

The behaviour of nitrogen and hydrogen can be understood on the basis of the above described mechanisms involving penetration of the protective scale by nitrogen and H_2/H_2O redox systems. Subsequent nitride formation, in suitable systems (e.g. pure chromium), and hydrogen dissolution can result.

Factors affecting penetration of protective scales

The above discussion has shown how, and under what conditions, a gas may penetrate a growing 'protective' oxide scale leading to compound formation at the scale–metal interface and in the metal.

Several factors are important in terms of how such penetration may affect the metal or alloy and are listed below.

(a) the solubility of the oxidising species in the metal phase,
(b) the free energy of formation of the corresponding compounds,
(c) whether a continuous compound layer or a dispersed phase is internally precipitated within the metal,
(d) the presence of liquid phases in the systems involved.

(*a*) *Solubility* The solubility of the oxidising species, Z, depends on its activity coefficient, γ_Z, in the solution and may be limited by the formation of compounds as explained below.

(*b*) *Free energy of formation of compounds* Assume that the compound forms according to

$$M + Z = MZ;\ \Delta G^{\ominus}_{15} \tag{6.15}$$

where M is any of the metallic elements in the alloy.

Then the solubility of Z in the alloy is limited to the mole fraction, X_Z, where

$$X_Z = \frac{1}{\gamma_Z}\frac{a_{MZ}}{a_M}\exp\left(\Delta G^{\ominus}_{15}/RT\right) \tag{6.16}$$

Generally, where MZ is a pure compound, this reduces to

$$X_Z = \frac{1}{\gamma_Z a_M}\exp\left(\Delta G^{\ominus}_{15}/RT\right) \tag{6.17}$$

Several compounds may form of course; the most stable compound forming first. As this compound forms, metal M is depleted in the alloy, the value of a_M falls, and X_Z correspondingly rises until the next most stable compound forms. This condition will arise at the scale–metal interface, the more stable compounds being found deeper within the alloy phase, and a mixture of compounds found closer to the metal surface.

(*c*) *Formation of a continuous layer or internal precipitation* If the content of metal M in the alloy is high enough, the compound may form a continuous layer at the metal–scale interface and thereby restrict further access to the metal by acting as a diffusion barrier.

This is most likely to be seen on pure metals such as when chromium is oxidised in air at 1200 °C giving a scale of Cr_2O_3, beneath which is formed a layer of Cr_2N. This complies with the above discussion if the nitride forms as a result of nitrogen penetration of the Cr_2O_3 scale.

Both internal attack and a layer of Cr_2N were produced when binary alloys of chromium containing Ti, Zr, Nb, etc., were oxidised in air at 1150 °C[12]. In addition to the Cr_2O_3 and Cr_2N layers, zones of internal nitridation were also formed.

The effect of oxide scale integrity was shown when cyclic oxidation of Ni–25Cr–6Al and Co–25Cr–6Al alloys resulted in internal precipitation of a zone of internal oxides forming closer to the metal–scale interface[13].

The formation of internal nitrides, carbides, etc., is particularly deleterious since the compounds formed are frequently those of the metals which are involved in the formation of protective oxide scales. When these elements are removed from a uniform distribution in the alloy, to be concentrated in a number of dispersed sites as precipitated compounds, it is not possible to form a continuous protective oxide layer on subsequent oxidation.

(*d*) *Formation of liquid phases* The formation of a liquid phase is particularly deleterious. Firstly, transport of both anions and cations is increased in the liquid phase. Secondly, if a continuous liquid phase is formed at the scale–metal interface, scale–metal adhesion is greatly impaired, if not completely lost. Thirdly, the liquid phase may not be a simple liquid compound but most likely a eutectic liquid. If the eutectic involves the parent metal of the alloy then severe grain boundary attack of the alloy is to be expected. This is because as the metal transports out through the liquid and scale layers it dissolves at the metal surface to maintain eutectic composition. The rate of dissolution is greatest at the grain boundaries which are, therefore, preferentially attacked with disastrous results for the mechanical properties of the metal or alloy. Figure 6.6 shows the results of exposing pure nickel to an atmosphere containing SO_2 at 1000 °C after forming a protective oxide layer in oxygen. Clearly, the sulphur has penetrated the oxide scale and appears to have formed a Ni–Ni_3S_2–NiO ternary eutectic which has resulted in extensive grain boundary attack of the metal. In this system a Ni–Ni_3S_2 eutectic forms at 637 °C, a ternary Ni–Ni_3S_2–NiO eutectic is not reported but the appearance of

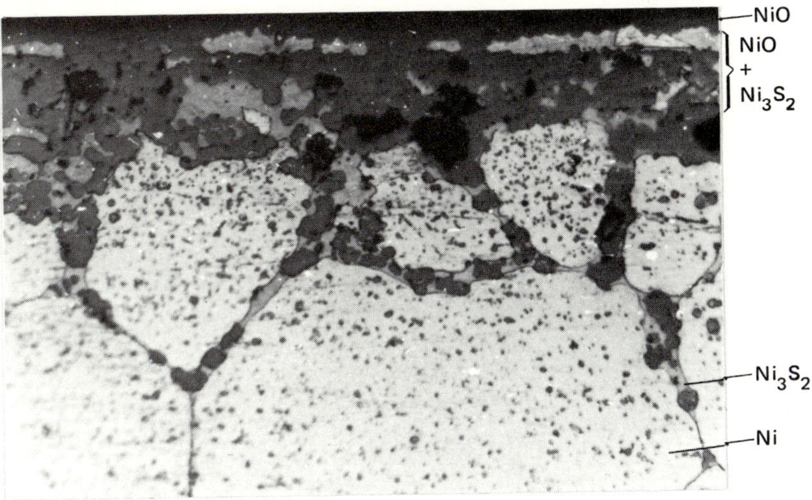

Fig. 6.6 Example of grain boundary penetration by eutectic liquid of the Ni–S–O system following penetration of a preformed oxide layer by SO_2 at 1000 °C. ×500

the oxide and sulphide phases in the nickel grain boundaries is strong evidence either for its existence or for appreciable oxide solubility in the $Ni–Ni_3S_2$ eutectic.

In addition to the penetration of scales by second oxidising elements, there is also the converse phenomenon of outward migration of cations different to those comprising the scale. A good example of this can be found in the oxidation of alloys containing chromium which forms a protective scale of Cr_2O_3. After oxidation for longer times, such scales are found to have outer layers enriched in iron, manganese, nickel, etc. These outer oxides usually exist as spinel phases $FeCr_2O_4$, $MnCr_2O_4$, or $NiCr_2O_3$ but, in the case of iron–chromium alloys, a layer of Fe_2O_3 can also develop as the outermost layer. Clearly cations, other than those of chromium, can migrate through Cr_2O_3, and Mn^{2+}, Fe^{2+}, and Co^{2+} cations appear to be the most mobile in descending order[14]. The apparent ease with which Mn diffuses through Cr_2O_3 scales accounts for the formation of MnS on the surface of Cr_2O_3 scales of alloys exposed to atmospheres containing oxygen and sulphur at low potentials, typical of coal conversion and oil reforming atmospheres.

Such outward migration of cations has been found to occur in Cr_2O_3 and Al_2O_3. There is also some evidence that the process is retarded when the initial oxide layer is produced in sulphur-free atmospheres which suggests that, when the oxide is grown in the presence of sulphur, the oxide structure is modified to accommodate the outward transport of cations through it.

The subject of penetration of protective oxide scales by cations and by second oxidants deserves more extensive investigation.

Both phenomena can be seen when Incoloy-800 is preoxidised in air to give a scale of Cr_2O_3 with a $MnCr_2O_4$ overgrowth and then, after exposure to a carburising Ar–10%CH_4 atmosphere, zones of internal carbide precipitation are found in the metal[15]. Thermal cycling, or simply cooling to room temperature, enhances the carburisation emphasising the physical nature of the mechanism by which the carburising gases penetrate the scale. Under normal circumstances the rate of penetration may be low, however the alloys involved are designed for continuous exposure to these conditions over a period of years.

Summary

In this chapter it has been shown how complex scales can form from complex atmospheres, under conditions such that a casual thermodynamic assessment would predict the formation of a scale of one component only. A correct understanding of the formation of duplex scales can only be achieved when factors relating to kinetics are considered together with thermodynamics. The second important point is that one of the most significant mechanistic steps in the breakdown of protective scales is the penetration of these scales by second oxidant species[8] or foreign cations[12]. These processes are being seen, increasingly, as essential steps in the initiation of failure of components which, although very resistant to oxidation in clean atmospheres, are sensitive to the presence of carbonaceous or sulphurous gas species in the atmosphere to which they are exposed.

References

1 Pettit, F. S., Goebel, J. A. and Goward, G. W., *Corros. Sci.*, **9,** 903, 1969
2 Birks, N., in: *High Temperature Gas–Metal Reactions in Mixed Environments*, ed. S. A. Jansson and Z. A. Foroulis, American Institute of Mining, Metallurgical, and Petroleum Engineers, New York, 1973; p. 322
3 Rahmel, A. and Gonzalez, J. A., *Werkst. u. Korrosion*, **21,** 925, 1970; also **22,** 283, 1971
4 Flatley, T. and Birks, N., *JISI*, **209,** 523, 1971
5 Wootton, M. R. and Birks, N., *Corrosion*, **12,** 829, 1972
6 Singh, P. and Birks, N., *Oxid. Metals*, **12,** 23, 1978
7 Luthra, K. L. and Worrell, W. L., *Met. Trans.*, **9A,** 1055, 1978

8 Birks, N., in: *Corrosion of High Temperature Alloys in Multicomponent Oxidative Environments*, ed. Z. A. Foroulis and F. S. Pettit, Electrochem. Soc., Las Vegas, 1976; p. 215

9 Pope, M. C. and Birks, N., *Oxid. Metals*, **12,** 191, 1978

10 Singh, P. and Birks, N., *Werkst. u. Korrosion*, **31,** 682, 1980

11 Rowlands, P. C. *et al.*, in: *Metal–Slag–Gas Reactions and Processes*, ed. Z. A. Foroulis and W. W. Smeltzer, Electrochem. Soc., Toronto, 1975; p. 409

12 Perkins, R. A., Corrosion in high temperature gasification environments, *3rd Annual Conference on Materials for Coal Conversion and Utilization*, DOE, October 1978

13 Giggins, C. S. and Pettit, F. S., *Oxid. Metals*, **14,** 363, 1980

14 Cox, M. G. C., McEnaney, B. and Scott, V. D., *Phil. Mag*, **26,** 839, 1972; also **29,** 595, 1974

15 Meier, G. H. and Perkins, R. A., *Oxid. Metals*, **17,** 235, 1982

7
Hot corrosion

Introduction

In addition to attack by reactive gases, alloys used in commercial environments, particularly those involving the combustion products of fossil fuels, undergo a mode of attack associated with the formation of a salt deposit, usually a sulphate, on the metal or oxide surface. This is called hot corrosion. The severity of this type of attack, which can be catastrophic, has been shown to be sensitive to a number of variables including deposit composition, atmosphere, temperature and temperature cycling, erosion, alloy composition, and alloy microstructure[1]. The purpose of this chapter is to introduce the reader to the mechanisms by which hot corrosion occurs. The examples used will be those associated with Na_2SO_4 deposits which are often encountered.

Once a deposit has formed on an alloy surface the extent to which it affects the corrosion resistance of the alloy will depend on whether or not the deposit melts, how adherent it is, and the extent to which it wets the surface. A liquid deposit is generally necessary for severe hot corrosion to occur although some examples exist where dense, thick, solid deposits have, apparently, resulted in considerable corrosion. The relationship of adhesion and wetting to hot corrosion has not been studied.

Once the alloy surface has been partially or completely wet by the molten salt, conditions for severe corrosion may develop. The hot corrosion of virtually all susceptible alloys is observed to occur in two stages: an initiation stage during which the rate of corrosion is slow and a propogation stage in which rapid, often catastrophic, corrosion occurs. This is illustrated for the commercial alloy IN-738 in Fig. 7.1. During the initiation stage, the alloy is being altered to make it susceptible to rapid attack. This alteration may include depletion of the element responsible for forming the protective scale (usually Al or Cr), formation of sulphides in the alloy due to sulphur penetration through the scale, dissolution of oxides into the salt, and development of growth stresses in the scale. This alteration also can result in shifts in salt composition toward more corrosive conditions. The length of the initiation stage varies from seconds to thousands of hours and depends on a large number of variables including alloy composition, alloy microstructure, salt composition, atmosphere, temperature, extent of thermal cycling, salt thickness, specimen geometry, and presence or absence of erosive conditions.

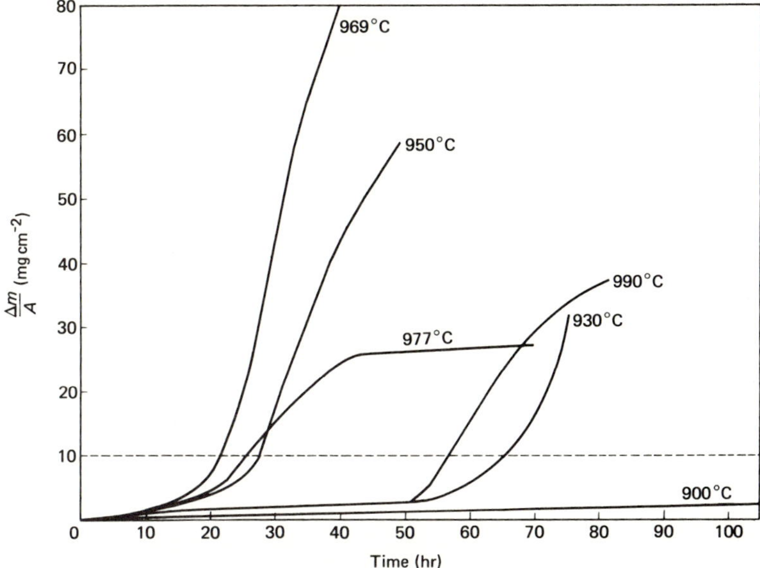

Fig. 7.1 Mass change versus time for IN-738 coated with 1 mg cm^{-2} Na_2SO_4 in 1 atm O_2

In many cases, the end of the initiation stage follows the local penetration of the salt through the scale and subsequent spreading along the scale–alloy interface. This situation, in which the salt reaches sites of low oxygen activity and is in contact with an alloy depleted in Al or Cr, generally leads to the rapid propagation stage. The propagation stage can proceed by several modes depending on the alloy and exposure conditions.

One group of propagation modes includes the salt fluxing reactions. Sodium sulphate may be considered to be composed of a basic component Na_2O (or O^{2-} ions) and an acid component SO_3. The salt fluxing reactions include basic fluxing, in which the oxide scale reacts with Na_2O and dissolves in the salt as anionic species, and acid fluxing, in which the scale reacts with SO_3 and dissolves as a cationic species. In a liquid sodium sulphate deposit, the sulphate ion decomposes according to

$$SO_4^{2-} = \tfrac{1}{2}O_2 + SO_2 + O^{2-} \tag{7.1}$$

A sodium sulphate melt therefore has an oxygen ion, or Na_2O, activity which is defined by the oxygen and sulphur dioxide potentials.

In *acid* fluxing, the oxygen ion concentration of the melt is low compared to the value required to maintain equilibrium in the dissociation reaction of the metal oxide according to

$$MO = M^{2+} + O^{2-} \tag{7.2}$$

Correspondingly, with low oxygen ion concentrations in the sulphate melt, acid fluxing can occur when the metal oxide simply dissolves into the melt according to equation 7.2.

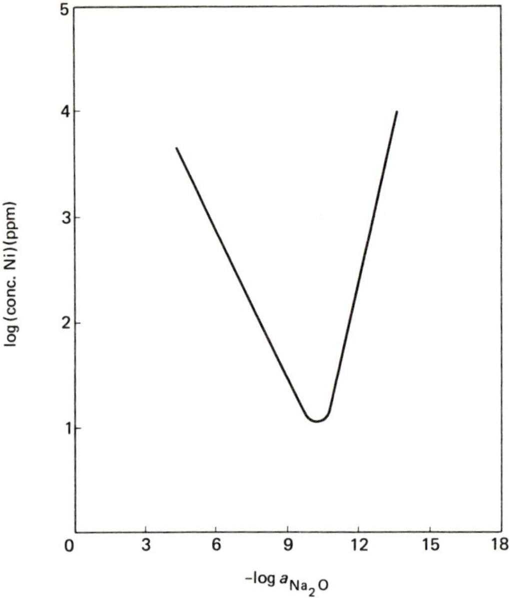

Fig. 7.2 Solubility of NiO in fused Na_2SO_4 at 1200 K (from Gupta and Rapp[2])

If the oxygen ion activity of the sulphate melt is high compared to that required to form complex anions according to equilibria such as

$$MO + O^{2-} = MO_2^{2-} \tag{7.3}$$

then complex ions can form and the metal oxide dissolves in the sulphate melt as a complex anion. This is known as *basic* fluxing. The two fluxing reactions are illustrated for NiO by the solubility plot[2] shown in Fig. 7.2. There is a minimum solubility at an activity of Na_2O of 10^{-10}. The solubility increases with increasing Na_2O activity (decreasing p_{SO_3}) according to the reaction

$$2NiO + O^{2-} + \tfrac{1}{2}O_2 \leftrightharpoons 2NiO_2^- \tag{7.4}$$

which corresponds to basic fluxing and increases with decreasing Na_2O activity according to the reaction

$$NiO \leftrightharpoons Ni^{2+} + O^{2-} \tag{7.5}$$

which corresponds to acid fluxing. Similar solubility behaviour occurs for other oxides.

Of course the stabilities of the various complex anions vary and conditions which may lead to the acid fluxing of one oxide could equally cause basic fluxing of another oxide. The terms acid and basic are relative and refer to the reaction that occurs rather than to just the condition of the melt, i.e. for each oxide there is a relationship such as in Fig. 7.2 for the condition of acid and basic fluxing. The curves for different oxides are displaced to left or right of each other depending on the relative stabilities of the compounds involved.

Basic fluxing

Figure 7.3 shows the rate of oxidation of Ni undergoing hot corrosion and the superposed micrographs indicate the reaction sequence. This sequence is shown schematically in Fig. 7.4.

A thin NiO scale forms on heating and is covered by the Na_2SO_4 as it melts. The continued formation of oxide rapidly lowers the p_{O_2} in the salt, as originally proposed by Goebel and Pettit[3], and the sulphur potential increases leading to transport through the thin oxide and sulphide formation at the scale–metal interface (Fig. 7.4(a)). The nature of the sulphur transport has not been verified but the fact that sulphides are observed in times as short as 10 seconds at 900 °C suggests that transport is not by lattice diffusion of sulphur ions through the NiO. Wagner *et al.*[4] found the diffusivity of S in NiO to be of the order of 10^{-14}–10^{-12} $cm^2\ s^{-1}$ depending on the oxygen partial pressure. If a value of 10^{-13} $cm^2\ s^{-1}$ is chosen the diffusion distance in 10 seconds at 900 °C is of the order of 10^{-6} cm which is small compared to the thickness of the initial NiO scale. Therefore, it appears a more likely mechanism is transport by SO_2 molecules penetrating through such defects in the scale as microcracks as was found by Wootton and Birks[5] for oxidation of Ni in Ar–SO_2 mixtures. The source of the SO_2 is the dissociation of the sulphate

$$SO_4^{2-} \rightarrow SO_2(g) + \tfrac{1}{2}O_2(g) + O^{2-} \qquad (7.6)$$

Also, as seen from equation 7.6, as SO_2 and O_2 are consumed the oxide ion activity in the salt will increase to maintain equilibrium resulting in the salt becoming more basic. The increase in basicity will be greatest over the areas where the sulphides form, i.e. where SO_2 is consumed most rapidly, and in these regions the NiO scale will react to form soluble nickelate ions in the melt according to equation 7.4 (Fig. 7.4(b)) which will diffuse to the salt–gas interface where the oxide ion concentration is low and will reprecipitate as NiO. The dissolution of the scale then allows the salt to penetrate it and spread along the scale–metal interface (Fig. 7.4(c)) lifting and cracking the scale. This cracking may actually be initiated by the formation of a Ni–S liquid phase at the scale–metal interface with a greater molar volume than that of the nickel. The cracking of the scale also allows oxygen penetration which oxidises the sulphides, thereby freeing sulphur to penetrate further into the

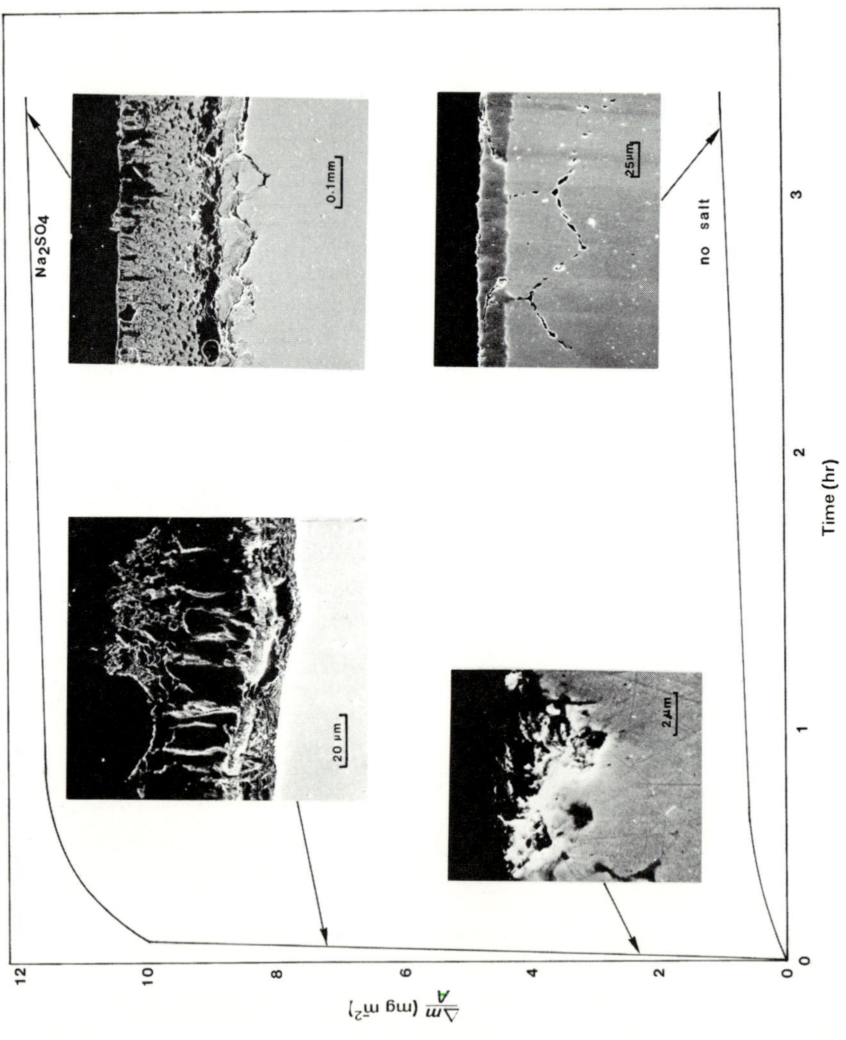

Fig. 7.3 Mass change versus time for Ni with and without Na_2SO_4 deposit

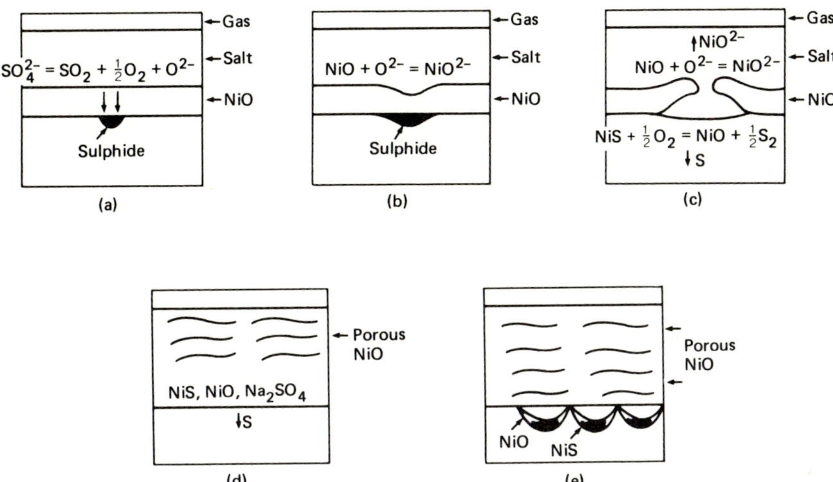

Fig. 7.4 Schematic diagram illustrating the Na_2SO_4 induced hot corrosion of pure nickel in 1 atm O_2

metal. The repetition of this process produces the porous, honeycomb-like NiO scale (Fig. 7.4(d)) and results in diffusion of sulphur and, subsequently, oxygen along the grain boundaries of the metal (Fig. 7.4(e)). Eventually, as Na_2SO_4 is trapped in the porous scale, the rapid reaction ceases and a dense, protective NiO layer forms. The reaction path in the Na_2SO_4 is superposed on the Ni–S–O stability diagram in Fig. 7.5.

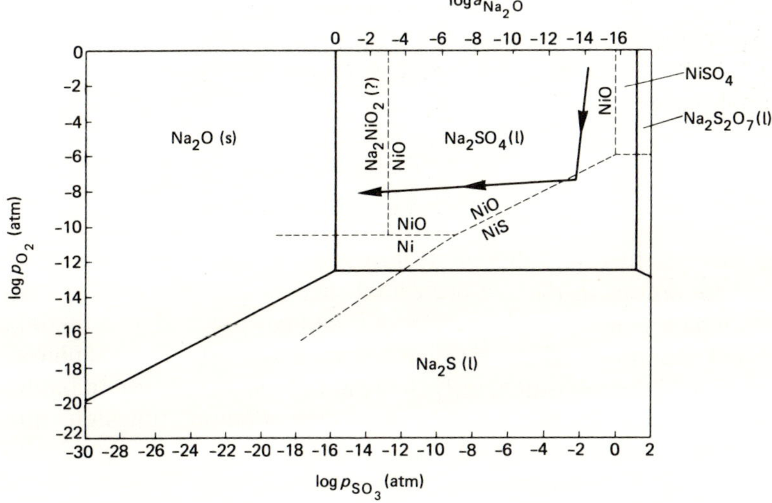

Fig. 7.5 Superposed stability diagrams for the Ni–S–O and Na–S–O systems showing schematic reaction path for the hot corrosion of Ni

A similar basic fluxing mode is also observed when the protective oxide is Cr_2O_3 or Al_2O_3 as in the hot corrosion of Ni–8Cr–6Al alloys[1]. Initial formation of Cr_2O_3 and Al_2O_3 depletes the salt melt of oxygen and lowers the oxygen potential. Accordingly, from equation 7.6, sulphate ions decompose further as $SO_4^{2-} \rightarrow SO_2 + \frac{1}{2}O_2 + O^{2-}$ and, at low oxygen potential, the sulphur activity increases leading to the formation of nickel sulphides. The low sulphur and oxygen potentials of the salt, resulting from the formation of oxides and sulphides, lead to an increase in the oxide ion, or Na_2O, activity in the melt.

In this way conditions are established for the basic fluxing of Cr_2O_3 and Al_2O_3 by the reactions

$$Cr_2O_3 + O^{2-} = 2CrO_2^{2-}$$

and

$$Al_2O_3 + O^{2-} = 2AlO_2^{2-}$$

forming chromate and aluminate ions, respectively, in solution in the molten salt. These ions migrate through the salt layer to sites of higher oxygen potential close to the salt–atmosphere interface where they precipitate out as Cr_2O_3 and Al_2O_3 releasing oxide ions according to

$$2CrO^{2-} = Cr_2O_3 + O^{2-}$$

and

$$2AlO_2^{2-} = Al_2O_3 + O^{2-}$$

The high oxygen potential at this position drives equation 7.6 in reverse such that the oxide ion, or Na_2O, activity becomes too low to support the existence of the complex anions which decompose as shown above. As in the case of pure Ni, the attack is not self-sustaining and gives over to sulphide formation in the alloy.

Acid fluxing

Acid fluxing may be further subdivided into alloy induced acid fluxing, in which the acid conditions in the salt are established by dissolution of species from the alloy which react strongly with Na_2O, and gas induced acid fluxing, in which the acid conditions are established by interaction with the gas phase.

Alloy induced acid fluxing is generally the result of the dissolution of oxides of the refractory metals in the Na_2SO_4 (e.g. molybdenum, tungsten, and vanadium forming molybdates, tungstates, and vanadates, respectively) thereby lowering the oxide ion concentration of the salt melt. This makes the salt much more acid and dissolution of oxides in the salt by dissociation according to equation 7.2 can occur. The commercial alloy B-1900 (Ni–8Cr–

6Al–6Mo–10Co–1.0Ti–4.3Ta–9.11C–0.15B–0.07Zr) is a particularly good example of this process. The significant difference between the oxidation and hot corrosion resistance of B-1900 has long been recognised. Figure 7.6 shows

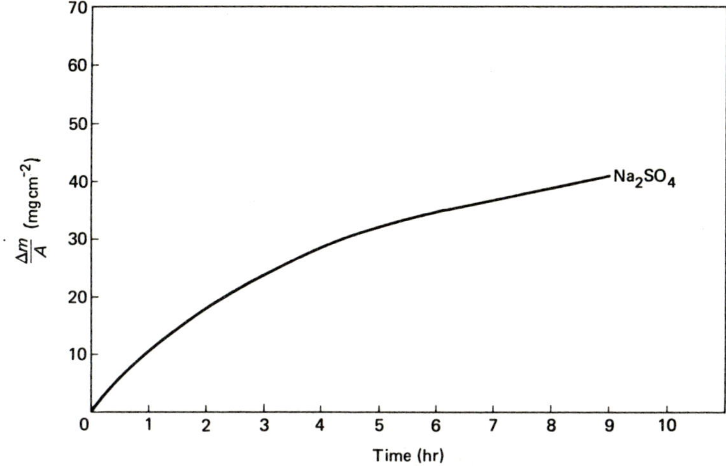

Fig. 7.6 Mass change versus time for B-1900 coated with 1 mg cm^{-2} Na_2SO_4 in 1 atm O_2 at 950 °C

the rate of oxidation of B-1900 at 950 °C in the presence of a 1 mg cm^{-2} Na_2SO_4 coating. (The rate of simple oxidation is too low to be observable on the scale of Fig. 7.6.) After a short incubation time, a rapidly accelerating mass gain occurs. The oxide scale formed on the specimens during the reaction is very porous and thick with a layered texture which peels off on cooling to room temperature. X-ray diffraction analysis of the spalled scale indicates a large amount of NiO and a small amount of $NiMoO_4$ and spinel. The oxide remaining on the specimens contains a large amount of $NiMoO_4$. Therefore, it is apparent the oxide peeled off just above a $NiMoO_4$ layer that was close to the alloy–oxide interface. The metallographic results from this region, shown in Fig. 7.7, indicate a region of Mo-rich phases restricted to a narrow zone between the outer porous scale and inner matrix. This layer contains Na, S, Ta, Mo, Ni, Cr, and Al. It appears this region may be a solution of Na_2SO_4 and Na_2MoO_4 into which Al_2O_3, Cr_2O_3, NiO, and Ta_2O_5 are dissolved. Beyond this zone, the layered scale contains mainly NiO. Internal oxides at the metal–scale interface consist mainly of Cr and Al oxides. It is clear that a continuous layer of Cr_2O_3 and Al_2O_3 is not developed on the alloy surface. Spherical chromium sulphide particles form in front of the internal oxide stringers.

Gas induced acid attack occurs when the atmosphere contains relatively high partial pressures of SO_3, the acid component of Na_2SO_4. Under these conditions, reaction 7.3 is forced to the left resulting in a salt melt of low

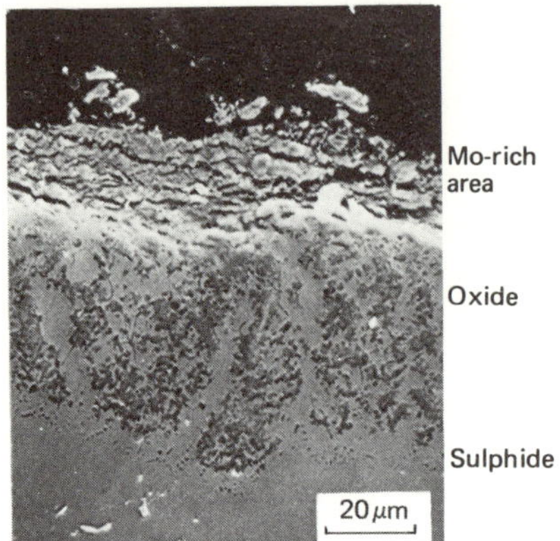

Fig. 7.7 The scale of B-1900 coated with Na_2SO_4 (1 mg cm^{-2}) and oxidised at 950 °C for 10 hours

oxide ion, or Na_2O, activity. This form of hot corrosion is prevalent at low temperatures between 600 and 800 °C. The exact mechanism of this form of corrosion is still somewhat uncertain, but the corrosion morphologies, which often are in the form of pits, have been well characterised.

Figure 7.8 shows the morphology of a pit formed in Co–18Cr–6Al exposed to O_2 with $p_{SO_3} = 2 \times 10^{-3}$ atm for 48 hours at 750 °C with a Na_2SO_4 coating. The pit is rich in Al and Cr with a Co-rich external layer and S-enrichment at the pit base. While the mechanism is not totally established, the following steps are known to occur. The reaction of CoO and SO_3 forms $CoSO_4$ which

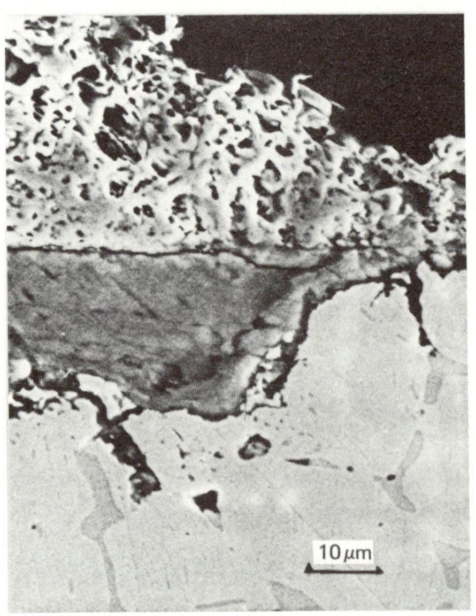

(a)

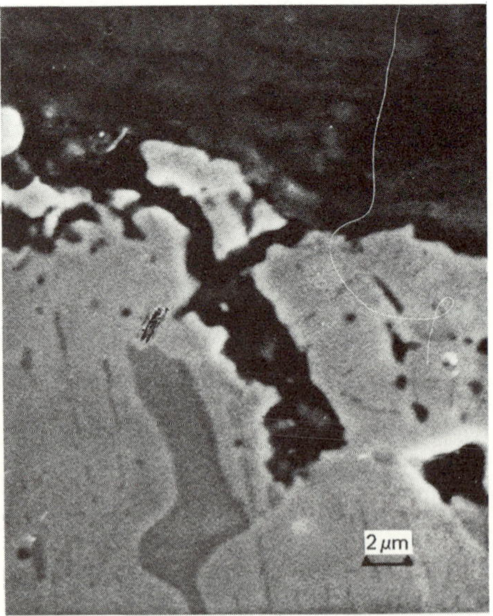

(b)

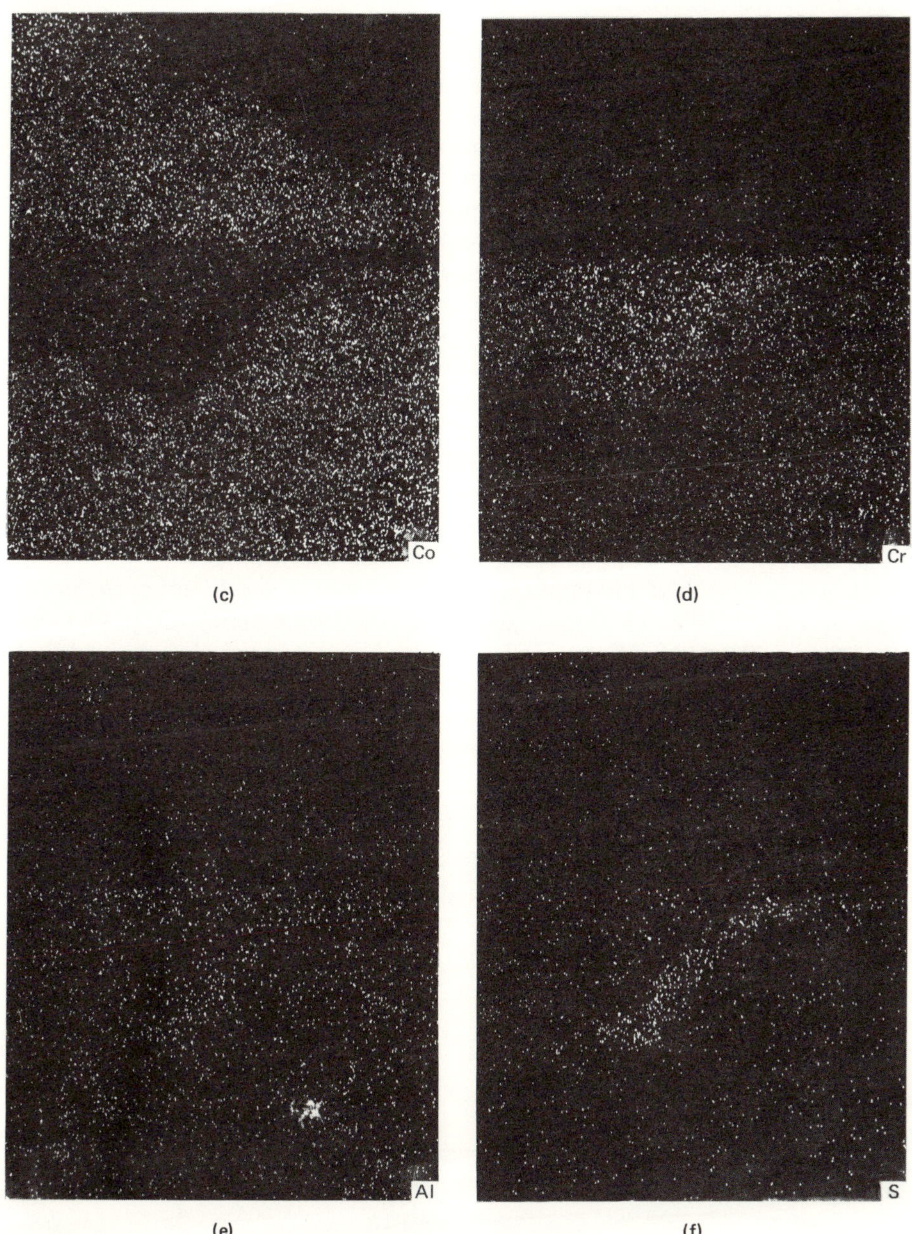

Fig. 7.8 Corrosion morphology for Na_2SO_4-coated Co–18Cr–6Al alloy exposed to $O_2 + SO_3$ ($p_{SO_3} = 2 \times 10^{-3}$ atm) for 48 hours at 750 °C. (a) and (b) scanning electron micrographs, (c), (d), (e), and (f) Co, Cr, Al, and S X-ray maps of (a)

forms a low melting point solution with Na_2SO_4. The liquid salt then penetrates the Al_2O_3 scale. Alternating layers of Al and Cr compounds are formed by selective removal of Al from the alloy. The Al-depleted alloy is converted to Cr_2O_3 to form the Cr-rich region and the remaining Co diffuses to the liquid–gas interface forming Co oxide or Co sulphate.

The propagation of hot corrosion may also proceed via reactions between the alloy and components induced from the salt. The components of greatest importance are sulphur, carbon, and chloride. The extensive sulphidation of the elements, such as Cr and Al, which are needed to form a protective oxide scale often results in the formation of non-protective oxide layers. This occurs even though the sulphides can redissolve to allow the formation of Cr_2O_3 and Al_2O_3 since the localisation of Cr and Al in the sulphides apparently alters the flux of these elements to the alloy–scale interface.

The presence of carbon, which can arise in some stages of combustion processes, often accelerates hot corrosion. The principal influence of carbon is in decreasing the oxygen pressure and increasing the sulphur pressure in the salt. This accelerates those mechanisms which are sensitive to these conditions, namely basic fluxing and sulphur induced corrosion.

The presence of NaCl in deposits can markedly accelerate hot corrosion as seen in Fig. 7.9. Here the corrosion of the commercial alloy FSX-414 (Co–29.5Cr–7W–10Ni–2Fe–0.35C) is seen to be slow in the absence of salt or the presence of Na_2SO_4 but accelerated by the presence of NaCl in the deposit.

Even though the absolute mass gains produced by the salt containing NaCl

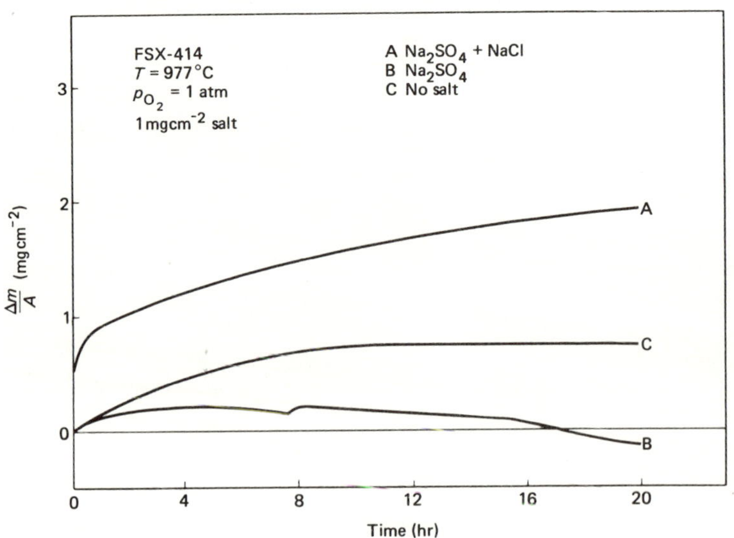

Fig. 7.9 Effect of deposit composition on the oxidation rate of FSX-414 at 977 °C

are not catastrophic, the useful thickness of the alloy is rapidly degraded. This degradation is in the form of internal sulphides and oxide-containing voids. The voids do not occur in the absence of NaCl (see Fig. 7.10).

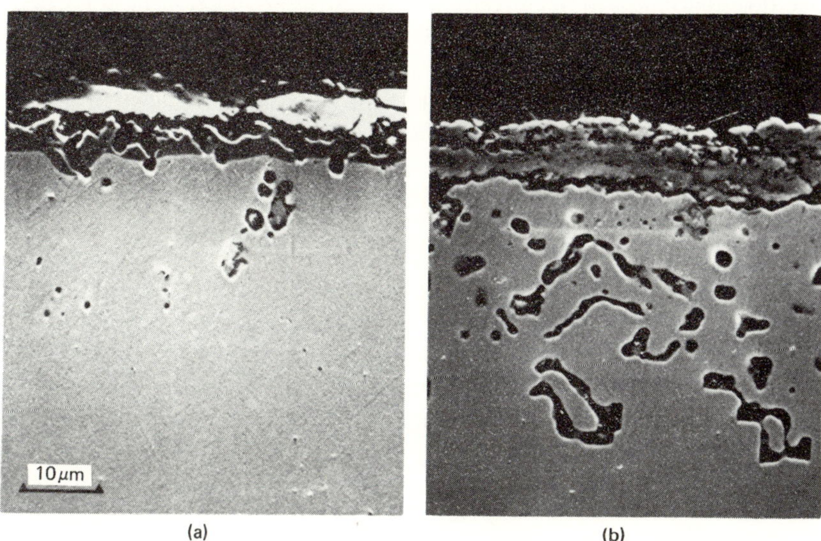

Fig. 7.10 Oxidation morphology for FSX-414 reacted at 977 °C in 1 atm O_2 with coatings of (a) Na_2SO_4 and (b) Na_2SO_4 + 20% NaCl

This morphology results from the penetration of Cl into the alloy to depths where the p_{O_2} is low and volatile chlorides of Cr can form. The formation of gaseous species results in the porosity and their subsequent oxidation, as they move toward the alloy–scale interface, results in the particulate oxide formation. The presence of NaCl also results in considerable scale spalling on cooling, as seen in Fig. 7.11. Similar results are obtained for both Cr_2O_3- and Al_2O_3-forming alloys of Co and Ni.

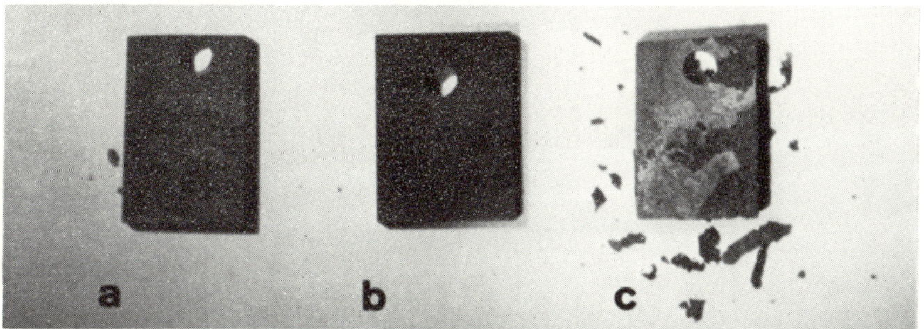

Fig. 7.11 Effects of Na_2SO_4 and NaCl on scale spalling for FSX-414: (a) no salt, (b) Na_2SO_4, and (c) Na_2SO_4 and NaCl

Summary

The above discussion has indicated corrosion morphologies typical of particular modes of attack. However, it should be recognised that many of these modes can occur in conjunction with one another, either sequentially at the same point or simultaneously at different points over the surfaces of alloys. The sequential basic fluxing and sulphidation has been observed for Ni–Cr–Al alloys[1] as well as pure Ni, initial basic fluxing is often observed to precede the acid fluxing of B-1900[1], and the interaction between sulphidation and acidic fluxing in the degradation of IN-738[6] has been observed. Also, it is clear from the above discussion that effects due to carbon and chloride can occur with most of the other attack modes. Therefore, in identifying hot corrosion mechanisms, it is important that the exposure conditions be extensively characterised. These conditions must also include alloy microstructure, thickness of salt deposit, thermal cycling, deposit composition, atmosphere composition, and temperature.

References

1 Giggins, C. S. and Pettit, F. S., Hot corrosion degradation of metals and alloys – a unified theory, *Final Report to Air Force Office of Scientific Research on Contract no. F44620-76-C-0123*, Pratt and Whitney Aircraft, June 1979
2 Gupta, D. K. and Rapp, R. A., *J. Electrochem. Soc.*, **127,** 2194, 1980
3 Goebel, J. A. and Pettit, F. S., *Met. Trans.*, **1,** 1943, 1970
4 Chang, D. R., Nemoto, R. and Wagner, J. B., Jr., *Met. Trans.*, **7A,** 803, 1976
5 Wootton, M. R. and Birks, N., *Corros. Sci.*, **12,** 829, 1972
6 Huang, T. T., *Ph.D. Thesis*, University of Pittsburgh, 1979

8
Atmosphere control for the protection of metals

Introduction

The exposure of metals to gases at high temperatures during production processes may be divided broadly under two headings

(a) reheating for subsequent working or shaping,
(b) reheating for final heat treatment, sometimes of finished components.

When reheating for subsequent working or shaping, the main concern is to heat the components as quickly and economically as possible to the working temperature within the metallurgical constraint of avoiding element redistribution and thermal cracking. In addition, it is usual to expose the metal directly to a burned fuel atmosphere containing about 1% excess oxygen necessary to ensure complete combustion and, therefore, the most economic use of the fuel. Under these conditions, as there is little likelihood of operating a controlled atmosphere policy, surface damage due to oxidation or scaling and, in the case of steels, decarburisation, must be removed at a later stage in the process. Indeed, if the surface of the material to be reheated already has some undesirable features, it may be possible to remove, or reduce, them by allowing some scaling to occur.

Alloys which are susceptible to surface deterioration during reheating may be protected either by controlling the atmosphere composition to minimise, or avoid, surface deterioration or by applying a coating to protect the alloy from the atmosphere.

Coatings are not an economic proposition in the treatment of large tonnage production in this way, but are applied to cause selective carburisation, etc., during surface hardening procedures. Alternatively, coating application is common during service and is a fairly successful way to avoid short-term damage to the underlying alloy.

This chapter is concerned with the use of gaseous atmospheres to control surface reactions during reheating for working or heat treatment.

The main application of controlled atmospheres is in the area of heat treatment of finished, machined, components or of articles of complex shape which cannot easily be treated subsequently for the removal of surface damage. In this context, atmosphere control can be considered under two headings: firstly where the treatment is aimed simply to prevent surface reaction and secondly where it is designed to cause a particular surface reaction to proceed such as carburising or nitriding. We are concerned only with the former.

Prevention or control of oxide layer formation

This is primarily a question of controlling the oxygen partial pressure of the atmosphere to a value low enough to prevent oxidation. Very simply, for a metal which undergoes oxidation according to the reaction

$$M + \tfrac{1}{2}O_2 = MO; \; \Delta G_1^{\ominus} \tag{8.1}$$

where MO is the lowest oxide of M, the oxygen partial pressure must be controlled so as not to exceed a value

$$(p_{O_2})_{M/MO} = \exp\,(2\Delta G_1^{\ominus}/RT) \tag{8.2}$$

Unfortunately $(p_{O_2})_{M/MO}$ is a function of temperature and has lower values at lower temperatures. Thus if an atmosphere is designed to be effective at high temperatures it may become oxidising as the temperature is reduced during cooling. A surface oxide layer may, therefore, form as the metal is cooled. Although the metallurgical damage to the surface will be negligible, the surface may be discoloured, i.e. not bright. This condition can be overcome to some extent by rapid cooling or by changing the atmosphere to a lower oxygen partial pressure just before, or during, cooling.

For alloys, it is clear that the most critical reaction must be considered when deciding on the composition of the atmosphere to be used. For this purpose, the activities of the alloy components must be known since, if the metal M in equation 8.1 exists at an activity a_M, the corresponding equilibrium oxygen partial pressure will be given by

$$(p_{O_2})_{a_M/MO} = \frac{1}{a_M^2}\exp\,(2\Delta G_1^{\ominus}/RT) \tag{8.3}$$

If the metal activities in the alloy are not known, then, by assuming the solution to be ideal, mole fractions may be used instead of activities to give a value of the oxygen partial pressure around which experiments must be performed to establish the correct atmosphere composition.

Figure 8.1 and Table 8.1 show values of the equilibrium oxygen partial pressures characteristic of copper, iron, nickel, and chromium and their oxides at various activities, derived using thermodynamic data from Kubaschewski *et al.*[1].

From Fig. 8.1 and Table 8.1, it is immediately obvious that the oxygen partial pressures are very low, especially in the case of chromium. The trend to lower oxygen partial pressures as the temperature is reduced is also clear. Finally, the effect of low metal activity, although it leads to higher oxygen partial pressures, is in fact relatively small.

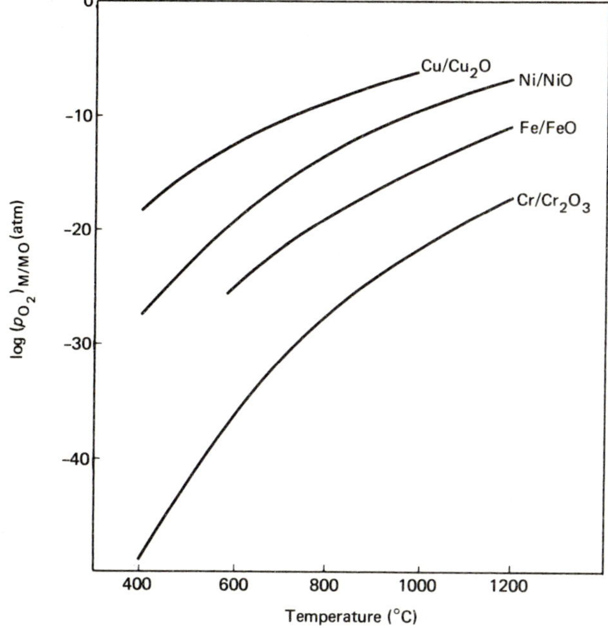

Fig. 8.1 Equilibrium oxygen partial pressures for several metal–oxide systems

Provision of protective atmospheres in the laboratory

Vacuum

To carry out heat treatment in vacuum is the first and perhaps the simplest method of avoiding surface oxide formation to be considered.

If the vacuum is against air, as is usual, it is possible to reduce the oxygen

Table 8.1 Approximate equilibrium oxygen partial pressures (atm)

Metal	Activity	400 °C	600 °C	800 °C	1000 °C	1200 °C
Cu	1	4.8×10^{-19}	3.4×10^{-13}	1.4×10^{-9}	4.4×10^{-7}	Liquid
	0.5	7.6×10^{-18}	5.6×10^{-12}	2.2×10^{-8}	7.0×10^{-6}	Liquid
Ni	1	1.7×10^{-28}	9.1×10^{-20}	2.7×10^{-14}	1.5×10^{-10}	8.4×10^{-8}
	0.5	6.8×10^{-28}	3.6×10^{-19}	1.1×10^{-13}	6.2×10^{-10}	3.4×10^{-7}
Fe	1	1.3×10^{-34}	2.4×10^{-25}	1.5×10^{-19}	1.5×10^{-15}	1.2×10^{-12}
	0.5	5.2×10^{-34}	1.8×10^{-24}	6.0×10^{-19}	6.0×10^{-15}	4.8×10^{-12}
	0.1	1.3×10^{-32}	2.4×10^{-23}	1.5×10^{-17}	1.5×10^{-13}	1.2×10^{-10}
Cr	1	7.9×10^{-50}	1.7×10^{-36}	3.8×10^{-28}	2.1×10^{-22}	3.1×10^{-18}
	0.1	3.7×10^{-49}	7.9×10^{-36}	1.8×10^{-27}	9.7×10^{-22}	1.4×10^{-17}
	0.01	1.7×10^{-48}	3.7×10^{-35}	8.2×10^{-27}	4.5×10^{-21}	6.7×10^{-17}

partial pressure to values as low as 10^{-10} or 10^{-11} atm. Comparing this with the values in Fig. 8.1, it is seen that this should be effective for copper above 670 °C, nickel above 950 °C, and iron above 1200 °C, but not at all for chromium. Due to the very slow reaction rates, however, it is generally possible to extend these limits to lower temperatures depending on the time of treatment.

With vacuum treatment there is always the possibility, especially at the higher temperatures, of the loss of volatile metals, such as manganese, from the surface regions of the component. For example, brass treated in this manner will undergo serious, if not total, loss of zinc.

Even with modern techniques, vacua of the above levels can only be achieved in relatively small chambers and the technique is largely restricted to laboratory use.

Gaseous atmospheres

Purified inert gases, mainly argon, can be used. However, due to the difficulty of reducing the oxygen partial pressure below 10^{-6} atm, such atmospheres rely for their effect largely on the reduced reaction rate at these reduced oxygen partial pressures. Inert atmospheres are also expensive to produce and are, therefore, largely restricted to the laboratory.

Low oxygen partial pressures can be provided and, more importantly, controlled by using 'redox' gas mixtures. These are mixtures of an oxidised and a reduced species which equilibrate with oxygen, e.g.

$$CO + \tfrac{1}{2}O_2 = CO_2; \Delta G_5^{\ominus} = -282\,200 + 86.7T\,\mathrm{J} \tag{8.4}$$

from which

$$p_{O_2} = \left(\frac{p_{CO_2}}{p_{CO}}\right)^2 \exp(2\Delta G_5^{\ominus}/RT)$$

or, more usefully

$$\frac{p_{CO_2}}{p_{CO}} = p_{O_2}^{1/2} \exp(-\Delta G_5^{\ominus}/RT) \tag{8.5}$$

Thus from equation 8.5 the ratio of carbon dioxide to carbon monoxide may be calculated for any oxygen partial pressure and temperature.

In Fig. 8.2 corresponding values of the ratio p_{CO_2}/p_{CO} and p_{O_2} are plotted for the temperatures 400, 600, 800, 1000, and 1200 °C. The lines corresponding to the equilibria between metal and oxide are also drawn in for Ni, Fe, and Cr. At log (p_{CO_2}/p_{CO}) values above one of these lines the metal will oxidise, below it the corresponding oxide will be reduced. Since the lines of log (p_{CO_2}/p_{CO}) versus log p_{O_2} are parallel, by calculating one point for any intermediate temperature the line for that temperature may be drawn in.

Similar, more useful, diagrams have been produced by Darken and Gurry[2]

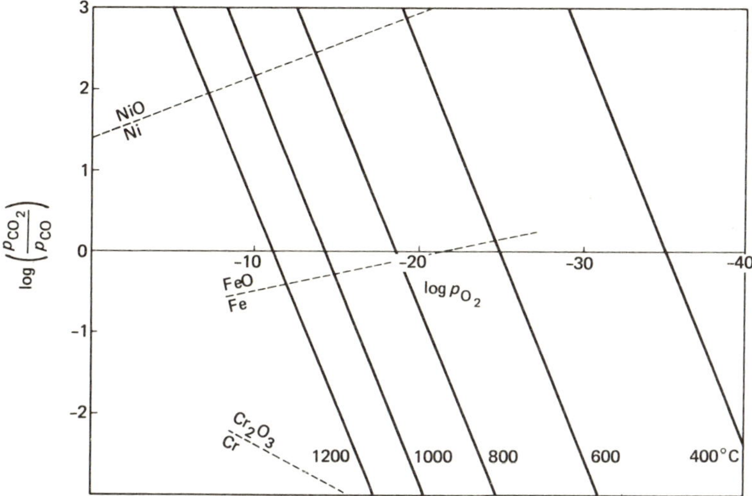

Fig. 8.2 Oxidising/reducing conditions in CO/CO_2 atmospheres

based on the modified Ellingham diagrams by Richardson and Jeffes[3] (see Chapter 2).

Several factors limit the use of CO_2/CO redox atmospheres, however. The usefulness of a redox gas system lies in its buffering capacity. Small concentrations of oxidising or reducing impurities, or leaks of oxygen into the furnace, are reacted with and removed, thus maintaining the desired oxygen potential, particularly in a flowing gas. This is obviously best achieved when the ratio is close to unity. Thus atmospheres which deviate from unity appreciably will lose, to some extent, their capacity for buffering. For example, in nickel, referring to Fig. 8.2, oxide formation will be prevented over the temperature range 400–1200 °C with CO_2/CO ratios varying between about 10^3 and 10^2 depending on temperature. This corresponds to a CO content of only 0.1 to 1%, thus protection of Ni is quite easy.

For iron, the CO_2/CO ratios vary over the same temperature range from about 3 to about 0.3. Such atmospheres are easily produced either by mixing the gases or, industrially, by partially burning fuel gas with air to produce the mixture. Furthermore, since the ratio is of the order of unity the mixtures provide good buffering action.

For chromium, the CO_2/CO ratios required are of the order of 2×10^{-4} at 1000 °C increasing towards lower temperatures. In such atmospheres only about 0.02% of CO_2 may be tolerated before oxidation occurs and such mixtures are expensive to produce and difficult to control. Even if production and control of these atmospheres could readily be achieved, a further problem arises in the form of the carbon activity, or carbon potential, of the gas mixture. Mixtures of CO_2 and CO will have a carbon potential by virtue of the reaction

$$2CO = CO_2 + C; \Delta G_7^{\ominus} = -170\,550 + 174.3T\,J \tag{8.6}$$

from which

$$a_C = \frac{p_{CO}^2}{p_{CO_2}} \exp(-\Delta G_7^{\ominus}/RT)$$

or

$$a_C = \frac{p_{CO}^2}{p_{CO_2}} \exp\left(\frac{20\,606}{T} - 21.06\right) \tag{8.7}$$

Protective atmospheres based on the CO_2/CO system therefore have a tendency to remove carbon from, or add carbon to, the metal depending on the composition of the metal to be treated. This is particularly important in the case of steels for which carbon is an essential constituent and the removal of carbon from the surface layers, decarburisation, is a deleterious reaction in terms of the strength of the steel surface.

In Fig. 8.3, carbon activities are shown as a function of p_{CO_2}/p_{CO} ratios for various temperatures, the corresponding metal–metal-oxide equilibria are superimposed. From Fig. 8.3 several conclusions may be drawn. At temperatures up to 1200 °C, the Cr/Cr_2O_3 equilibrium corresponds to carbon activities

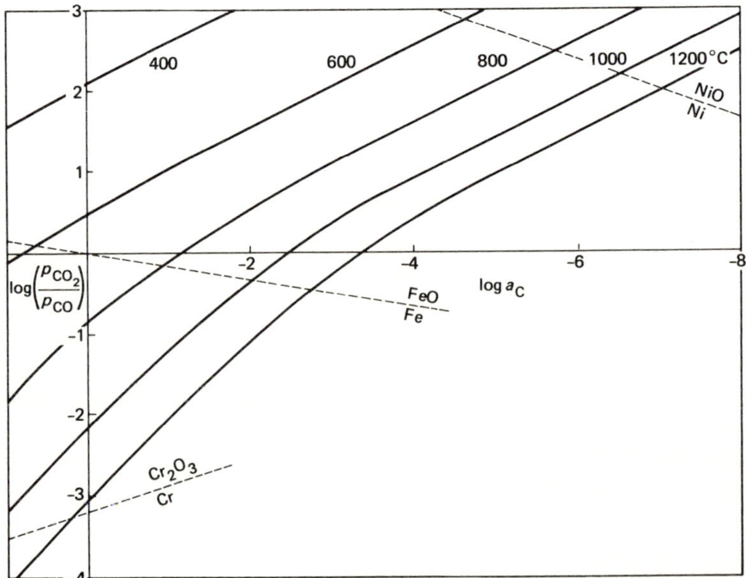

Fig. 8.3 Carbon activities in CO/CO_2 atmospheres

in excess of unity, therefore treatment of Cr in these atmospheres will lead to 'sooting' in the furnace and to severe carburisation of the metal. Such atmospheres are, therefore, unsuitable for the complete protection of chromium. For iron, the CO_2/CO ratios required for the prevention of scaling are easily produced and maintained being within the range 0.1 to 10 for all common temperatures. The problem here is that, during heating of steels, the above CO_2/CO ratios correspond to low carbon activities and therefore lead to decarburisation. Consequently, in order to avoid decarburisation, it is necessary to use CO_2/CO ratios much lower than those required to protect the iron from scaling. For example, using Fig. 8.2 at 1000 °C, the value of $\log(p_{CO_2}/p_{CO})$ required to protect iron is −0.3, corresponding to a CO_2/CO ratio of 0.50, but this corresponds to a carbon activity of 0.01 or a carbon content of 0.021 wt% according to the relationship of Ellis *et al.*[4] between carbon activity, a_C, carbon mol fraction, X_C, and temperature, T, for plain carbon steels in the austenite phase field

$$\log a_C = \log[X_C/(1-5X_C)] + 2080/T - 0.64 \tag{8.8}$$

Thus, if the steel has a carbon content of 0.8% which corresponds to a mol fraction of 0.036 and, from equation 8.8, to a carbon activity of 0.433, the atmosphere composition required to prevent decarburisation is obtained by locating log 0.433 = −0.364 and reading the corresponding $\log(p_{CO_2}/p_{CO}) = -1.75$ on the line for 1000 °C. This corresponds to an atmosphere of CO_2/CO ratio of 0.0178 or, since $p_{CO} + p_{CO_2} = 1$, to a composition of 98.25% CO and 1.75% CO_2. Thus, in order to protect a plain carbon steel containing 0.8% C, the above atmosphere should be used. If the atmosphere contains more CO_2, then some decarburisation will occur, although it may contain up to 50% CO_2 before the iron begins to oxidise.

Similar calculations may be carried out for nickel although, from Fig. 8.3, it is obvious that this metal will be protected by atmospheres of CO_2 containing only 1% CO or less, depending on temperature.

It is clear that, in order to protect alloys containing chromium from oxidation, atmospheres based on the CO_2–CO system cannot be used if carburisation is to be avoided. Here, and in other cases where strong carbide formers are involved, protection may be achieved by using the H_2–H_2O system for which the relevant equilibrium is

$$H_2 + \tfrac{1}{2}O_2 = H_2O; \ \Delta G_9^\ominus = -247\,000 + 55T \text{ J} \tag{8.9}$$

for which

$$\frac{p_{H_2O}}{p_{H_2}} = p_{O_2}^{1/2} \exp(-\Delta G_9^\ominus/RT)$$

or

$$\frac{p_{H_2O}}{p_{H_2}} = p_{O_2}^{1/2} \exp\left(\frac{29\,844}{T} - 6.65\right) \tag{8.10}$$

From equation 8.10, the ratio of water vapour to hydrogen in the atmosphere may be calculated for any oxygen partial pressure and temperature.

Protective atmospheres based solely on the H_2O/H_2 gas system are hardly used outside the laboratory. As may be seen from Fig. 8.4, however, both

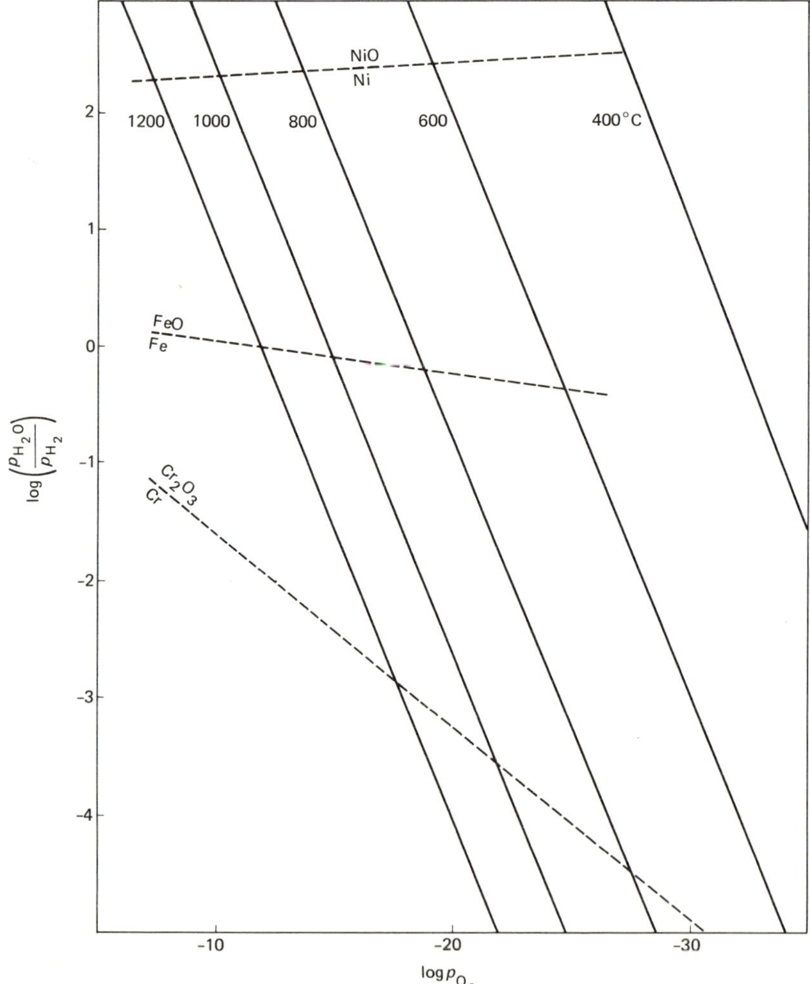

Fig. 8.4 Oxidising/reducing conditions in H_2/H_2O atmospheres

nickel and iron can be protected at all temperatures in hydrogen containing 10% H_2O and so, for laboratory use, cylinder hydrogen, usually containing a few parts per million of water vapour, can be used without further purification.

The system is of most use in the protection of alloys containing high chromium concentrations, such as stainless steels for which the required water content is as low as, or less than, 1000 ppm, depending on temperature. The advantage of this system is that the water vapour content can be controlled by cooling the gas to a low temperature thereby removing water from the gas by condensation. Similarly the water content of the gas may be monitored and, using feedback, controlled by dew-point measurement.

To heat stainless steels for softening, it is also possible to dilute the redox gas with an inert gas, such as nitrogen for which use is made of the relatively cheaply available ammonia which can be cracked according to equation 8.11.

$$2NH_3 = 3H_2 + N_2 \tag{8.11}$$

By catalysing this reaction at 800 °C, almost complete decomposition is achieved. This is important if excessive nitrogen pick-up is to be avoided as this occurs most readily from uncracked NH_3 molecules.

Suppose it is required to bright anneal a stainless steel containing 20% Cr, 10% Ni, and 0.05% C at 1000 °C in a cracked ammonia atmosphere. The maximum dew-point of the gas may be calculated as follows.

$$2Cr + 3H_2O = Cr_2O_3 + 3H_2; \Delta G^\ominus_{12} = -378\,300 + 94.5T \text{ J} \tag{8.12}$$

Thus

$$\frac{a_{Cr_2O_3}}{a^2_{Cr}}\left(\frac{p_{H_2}}{p_{H_2O}}\right)^3 = \exp\left(\frac{378\,300 - 94.5T}{RT}\right)$$

$$= 4.31 \times 10^{-10} \text{ at } 1000\,°C \tag{8.13}$$

and

$$\frac{p_{H_2}}{p_{H_2O}} = a^{2/3}_{Cr} \times 3.51 \times 10^3$$

In these alloys the activity coefficient of chromium is close to two, also the mol fraction of Cr corresponding to 20% in iron is $X_{Cr} = 0.19$. Thus $a_{Cr} = 0.35$.

The required ratio, p_{H_2}/p_{H_2O}, is therefore

$$\frac{p_{H_2}}{p_{H_2O}} = 0.35^{2/3} \times 3.51 \times 10^3 = 1.84 \times 10^3 \tag{8.14}$$

In ammonia, cracked according to equation 8.11, there are three parts hydrogen and one part nitrogen, thus the hydrogen partial pressure in this atmosphere is 0.75 atm. Therefore, from equation 8.14, we have

$$p_{H_2O} = 0.75/(1.84 \times 10^3) = 4.08 \times 10^{-4} \text{ atm}$$

The dew-point of the gas mixture must therefore be controlled to produce a water vapour partial pressure no greater than 4.08×10^{-4} atm. The relevant dew-point may be obtained from steam tables (−29 °C) or may be calculated as follows.

The Clausius–Clapeyron equation relates the saturated vapour pressure, P, to temperature and the latent heat of evaporation, L, as

$$2.303\frac{\partial \log P}{\partial(1/T)} = -\frac{L}{2.303R} \qquad (8.15)$$

Integrating this between T_1 and T_2 we have

$$\log\frac{P_1}{P_2} = -\frac{L}{2.303R}\left(\frac{1}{T_1} - \frac{1}{T_2}\right)$$

For water $P_2 = 1$ atm at 373 K and $L = 41\,000$ J mol^{-1}. Thus

$$\log P_1 = -\frac{41\,000}{2.303R}\left(\frac{1}{T_1} - \frac{1}{373}\right) \qquad (8.16)$$

The required dew-point is the temperature at which $P_1 = 4.08 \times 10^{-4}$ atm, i.e. 234.3 K (or −38.7 °C) according to equation 8.16. Correspondingly the dew-points of such atmospheres, used for bright annealing of stainless steels, would be controlled to about −40 °C.

Even with such atmospheres, however, there may be complications due to hydrogen uptake, decarburisation, and nitriding. Hydrogen uptake and nitriding in cracked ammonia atmospheres are both avoided largely by ensuring that the time of treatment is the minimum for the metallurgical requirements. In addition, by restricting the treatment to items of thin section, the heating time is reduced and any hydrogen dissolved is able to diffuse out quickly. Decarburisation may occur by the following reaction.

$$C + H_2O = CO + H_2; \ \Delta G^{\ominus}_{17} = 135\,400 - 1426T \text{ J} \qquad (8.17)$$

Reaction 8.17 proceeds rapidly and is therefore dangerous. Fortunately, the most usual application of the cracked ammonia atmosphere is to stainless steels where a low carbon content is an advantage.

Provision of controlled atmospheres in industry

When a process is considered for use in industry it must satisfy certain criteria. It must be

(a) effective, i.e. achieve the technical objective,
(b) reliable, i.e. capable of monitoring and control,
(c) economical, i.e. basically inexpensive to install and use and capable of accepting full production throughput.

For most applications the above requirements rule out systems based on high vacuum or inert gases, and the general practice is to use atmospheres derived from fuels. The gases used are therefore mixtures of N_2, CO, CO_2, H_2, H_2O, and CH_4 which make up the products of combustion of fuels. Recently, atmospheres based on nitrogen have been used at an increasing rate.

In the early days of atmosphere control, town gas was used for the preparation of atmospheres, however, being derived from coal, this inevitably involved the need for a plant for sulphur removal and variation in analysis made consistent control of the atmosphere composition difficult.

Recently, the availability of liquid petroleum gas (LPG), e.g. propane, with reproducible analysis and low sulphur content has improved the situation together with the use of natural gas for the same purpose.

It will also be seen that the large quantities of high purity nitrogen, produced when producing liquid oxygen for steelmaking, represent a convenient feedstuff for controlled atmospheres.

Types of atmosphere

Starting from the fuel and air, various types of atmosphere can be produced. The main, or common, differentiation is between 'exothermic' and 'endothermic' atmospheres. The nomenclature is ambiguous and it is as well to be clear about its meaning.

An exothermic atmosphere is produced *exothermically* by burning the fuel with measured amounts of air. This type of atmosphere has the highest oxygen potential.

An endothermic atmosphere is produced by heating, by external means, a mixture of fuel gas with air over a catalyst to provide a gas containing reducing species. This atmosphere has a low oxygen potential and heat is *absorbed* during its preparation, hence the atmosphere is described as *endothermic*.

Exothermic atmospheres

The first stage of production is a combustion chamber where gas and air are reacted, the mixture being adjusted between limits, capable of supporting combustion, described as 'rich' or 'lean' depending on whether the gas has low or high concentrations of oxidised species, respectively.

From the combustion chamber, the gas produced is cooled by water spray which removes sulphurous gases and reduces the water content usually to a dew-point of about 35 °C, equivalent to a water vapour content of around 6.5% by volume.

Typical rich exothermic and lean exothermic compositions are given, in percentages, as

Rich:	$5CO_2$	$9CO$	$9H_2$	$0.2CH_4$	$7H_2O$	$70N_2$
Lean:	$10CO_2$	$0.5CO$	$0.5H_2$	—	$7H_2O$	$82N_2$.

The rich atmosphere can be used for the bright annealing of low carbon steels.

The atmospheres can be refined further by reducing the water vapour content, this can be accomplished by refrigeration causing the water to condense. The temperature of this treatment is normally restricted to 5 °C, otherwise ice may form and block the process whereas water can easily be drained away. More effective drying down to dew-points in the region of −40 °C can be achieved using activated alumina or silica gel towers. Used in pairs, normally one absorber is operational whilst the other is regenerated.

These treatments lead to dew-points of about 4 °C (0.8% H_2O) and −40 °C (0.04% H_2O) and, apart from the reduced water vapour content, these atmospheres are identical with those given above and could be referred to as 'rich dried' or 'lean dried' atmospheres. Such atmospheres simply present a way of slightly reducing the oxygen potential, since, at temperature, the reaction

$$CO_2 + H_2 = H_2O + CO \tag{8.19}$$

will tend to equilibrate. Thus the dew-point of the cold, dried gas cannot be maintained but the total oxygen content and the oxygen potential are reduced.

A really protective gas, or even reducing gas, can be obtained if, after drying, the gas is 'stripped' of CO_2. This can be achieved in several ways. High pressure water scrubbing will remove some CO_2 but the process is not widely used. More usual are methods of absorbing CO_2 either chemically, using mono-ethanolamine (MEA) aqueous solution, or physically, using a molecular sieve.

MEA method

In this technique the gas is passed counter-current to a 15% solution of mono-ethanolamine in water and CO_2 is absorbed down to about 0.1%, sulphurous gases are also absorbed. The used MEA solution is regenerated by heating when the CO_2 and other absorbed gases are driven off. The heat required for this is frequently obtained by combining the regenerator and combustor in one unit.

After stripping the CO_2, the gas must now be dried, usually using silica gel or activated alumina.

Molecular sieve method

The molecular sieves used are artificial zeolites which adsorb both carbon dioxide and water vapour on the surface of structural cavities of molecular size. The zeolite can be regenerated and the method has the advantage of supplying stripped *and* dried gas in one operation.

These synthetic zeolites are also used in twin towers; both CO_2 and H_2O are removed, by adsorption, down to 0.05% by volume in both cases. Alternative adsorption and regeneration cycles are used, as reactivation of modern zeolites can be carried out at room temperature using a CO_2- and H_2O-free purge gas.

Typical dried and stripped exothermic gases have analyses such as

$$0.05\%\ CO_2,\ 0.5\%\ CO,\ 0.5\%\ H_2,\ 0\%\ CH_4,\ 0.04\%\ H_2O,\ 99\%\ N_2$$

which may be applied to the bright annealing of carbon steels.

Hydrogen-enriched gases

In order to reduce the oxygen potential of the above gas still further it is necessary to increase the hydrogen content. This can be done by producing an initially richer gas by reducing the air-to-fuel ratio in the combustor and so, after drying and stripping, the gas contains extra H_2 and CO. Alternatively, steam may be added to the stripped gas when, on reheating, the reaction $H_2O + CO = H_2 + CO_2$ occurs over a catalyst. The H_2O and CO_2 are once more removed as described above.

Hydrogen may also be added in the form of ammonia which will crack over a catalyst. A typical dried, stripped, hydrogen-enriched atmosphere would have the following composition.

$$0.05\%\ CO_2,\ 0.05\%\ CO,\ 3\text{–}10\%\ H_2,\ 0\%\ CH_4,\ 0\%\ H_2O,\ 90\text{–}97\%\ N_2$$

Such an atmosphere would be quite suitable for the bright annealing of carbon steel or for stainless steel, having a dew-point of about −40 °C.

Endothermic atmospheres

These are produced by reacting mixtures of fuel and air that are not capable of supporting combustion. Usually the aim is to crack the hydrocarbons over a catalyst at about 1050 °C and convert them mainly to CO and H_2. If the mixture is too rich, deposition of carbon could occur in the catalysis chamber and so the mixture is adjusted to avoid this.

Steam can also be used as the oxidant as partial, or total, replacement for air, in which case a hydrogen-enriched gas is produced of composition typically

$$5\%\ CO_2,\ 17\%\ CO,\ 0.4\%\ CH_4,\ 0.5\%\ H_2O,\ 0.0\%\ N_2,\ 71\%\ H_2$$

Further addition of steam will, on passing once more over a catalyst, push the equilibrium ($H_2O + CO = H_2 + CO_2$) to the right and, after drying and stripping of CO_2, produce compositions such as

$$0\text{–}0.5\%\ CO_2,\ 0\text{–}5\%\ CO,\ 75\text{–}100\%\ H_2,\ 0\text{–}0.5\%\ CH_4,\ 0\%\ H_2O,\ 0\text{–}25\%\ N_2$$

depending on the air/steam mixture. Such gases are quite suitable for the bright annealing of stainless steel due to the low oxygen and carbon potentials.

Nitrogen-based atmospheres

A glance at the analysis of dried, stripped, exothermic atmospheres will confirm that they are predominantly pure nitrogen. Basically the fuel has been used to remove oxygen from the air.

An alternative method would be to liquefy the air and fractionate it to produce pure nitrogen. This is, already, a common process carried out to produce oxygen for bulk steelmaking purposes. The nitrogen that is a convenient by-product will be available in considerable volume at steel-making sites that have on-plant oxygen facilities. Using vacuum insulated transporters, liquid nitrogen can be delivered over long distances and is being used in increasing quantities in bright treatment facilities.

The typical analysis of fractionated nitrogen is 99.99% N_2, 8 ppm O_2, dew-point −76 °C. Due to the method of production the analysis can be guaranteed. Since there is no reducing gas species present, there is no fire or explosion hazard. Neither is the gas toxic although, of course, it will not support life.

Even where reducing gases must be used for carburising, for example, a good precaution is to have a standby nitrogen connection to the furnace for emergency flushing. This may be necessary during hazardous conditions following failure of the gas power generator or water supply. Not only will a nitrogen flush under such conditions avoid a possible catastrophe but the work will be protected during cooling. In addition, there is no pollution.

It is also possible to mix nitrogen with controlled amounts of hydrogen, hydrocarbons, and/or ammonia to produce atmospheres capable of most, if not all, controlled atmosphere treatments. The only exceptions arise when metals containing strong nitride formers are to be treated, in which case a similar atmosphere based on argon is to be preferred. The actual choice will depend on the duration of the treatment compared with the rate of reaction with nitrogen gas. In fact, the rate of uptake of nitrogen from N_2 gas is much slower than when ammonia is present as the uncracked molecular species.

The economics of using nitrogen as a controlled atmosphere source become more attractive when such factors as improved safety, reliability, and, therefore, productivity are considered[5]. Furthermore, the present tendency is to move away from oil towards electricity in which case nitrogen atmospheres will be particularly attractive.

Monitoring and control

For most applications, the analysis can be defined in terms of CO_2 and H_2O content and these can be measured conveniently, but not continuously, using chemical absorption techniques. Once the correct CO_2 and H_2O levels have been established it is usually simply a matter of controlling the feeds of fuel and air to the appropriate levels.

For CO_2 monitoring on a more sophisticated level, infra-red absorption can be used; H_2O can also be monitored using this technique.

A convenient method of continuously monitoring the gas for H_2O content is to measure the dew-point using a commercially available dew-point meter. Certain meters, working on a conductance or capacitance principle, can be set to feedback to control the input, or drying, parameters, or to trigger alarms if a certain limiting value is exceeded.

Whereas the above methods allow the oxygen potential of an atmosphere to be deduced from measurements of the CO_2 and H_2O contents, a more recent development allows the oxygen potential, or partial pressure, to be measured directly and continuously. This technique is based on the use of a high temperature galvanic cell. The instrument is simply a tube of zirconia stabilised with CaO to retain the high temperature cubic structure. Within the tube an oxygen partial pressure is imposed either by flushing with a gas of known oxygen potential or by using a metal–metal-oxide mixture. A platinum wire establishes contact with the inner surface of the tube and a second wire is wound in intimate contact with the outside of the tube. The tube is then placed in the furnace atmosphere at temperature. Since the stabilised zirconia has conductivity for oxygen ions only, the following galvanic cell is established.

$$\ominus \quad Pt/p'_{O_2} \;\Big|\; \underset{\longleftarrow O^{2-}}{ZrO_2} \;\Big|\; p''_{O_2}/Pt \quad \oplus$$

The free energy change for this reaction is given by

$$\Delta G = RT\ln(p'_{O_2}/p''_{O_2}) \tag{8.20}$$

Consequently the emf of the cell is given by

$$E = RT/4F\ln(p''_{O_2}/p'_{O_2}) \tag{8.21}$$

In practice, the reference electrode could be controlled by flushing air or by establishing a reference oxygen partial pressure by a mixture of a suitable metal and its oxide such as Ni and NiO. Clearly the emf of such a cell can be indicated continuously and also used for feedback to correct the feed settings of fuel and air.

Difficulties encountered include thermal shock requiring relatively slow heating and cooling rates, although if mounted permanently in the furnace these should be achieved. It is important to mount the cell in a position that is exposed to the same temperature and gas composition as the working volume of the furnace to avoid unrepresentative readings. In addition, the cell will only indicate the equilibrium oxygen partial pressure, regardless of whether or not the atmosphere has equilibrated. This is because the gas will equilibrate locally on the platinum contacts of the cell.

Heating methods

It is clear that an independent source of heating is essential. Electrical heating is clean, controllable, and simple. Alternatively, gas can be burned within incandescent tubes built into the furnace walls heating by radiation. With atmospheres of high carbon potential, the possibility of carburising the electrical heating elements must also be considered and suitable shielding arranged.

References

1 Kubaschewski, O., Evans, E. H. and Alcock, C. B., *Metallurgical Thermochemistry*, Pergamon Press, Oxford, 1967
2 Darken, L. S. and Gurry, R. W., *Physical Chemistry of Metals*, McGraw-Hill, New York, 1953
3 Richardson, F. D. and Jeffes, J. H. E., *JISI*, **160,** 261, 1948
4 Ellis, T., Davidson, S. and Bodworth, C., *JISI*, **201,** 582, 1963
5 Bowes, R. G., *Heat Treatment of Metals*, **4,** 117, 1975

Further reading

1 Fairbank, L. H. and Palethorpe, L. G. W., Heat treatment of metals, *ISI Special Report no 95*, 1966; p. 57–69
2 Vaughan, C. A. and Zeigler, J. G., *Industrial Heating*, September, 1976; p. 8
3 Record, R. G. H., *Metals and Materials*, January, 1977; p. 25
4 Cutts, R. V. and Dan, H., *Inst. of Iron and Steel Wire Manufacturers Conferences*, Harrogate, March 1972; Paper 7

9
Decarburisation of steels

Introduction

The tensile strength of a heat treated steel depends primarily on the carbon content. The maximum stress on a component in bending occurs at the surface. Clearly, if excessive failure of steel components in service, particularly under reversing bending stresses, is to be avoided, the specified carbon content must be maintained within the surface layers of the component. This is particularly necessary with rotating shafts and stressed threads in bolts, etc.

Unfortunately, there is a strong tendency for carbon to be lost from the surface of steel when it is reheated for hot working or for heat treatment. This loss of carbon, decarburisation, is one of the oldest, most persistent, problems in ferrous production metallurgy.

During normal reheating, which is usually carried out in a large furnace where the steel stock is in contact with the products of combustion of a fuel (usually oil or gas), the steel also oxidises to form a scale and, therefore, undergoes simultaneous decarburisation and scaling.

Reheating for heat treatment to impart the final mechanical properties may be carried out under a controlled atmosphere to avoid carbon loss from the surfaces of a finished component.

The mechanism by which the decarburisation of steels occurs is well understood[1,2], particularly in the case of plain carbon and low alloy steels to which the following discussion is confined.

Basically, in the oxidising atmosphere of the furnace, a scale of iron oxides forms and grows. At the scale–metal interface, carbon interacts with the scale to form carbon monoxide by the reaction

$$\underline{C} + FeO = Fe + CO; \Delta G^{\ominus} = 147\,763 - 150.07T \text{ J} \tag{9.1}$$

This reaction can only proceed if the carbon monoxide reaction product can escape through the scale. In general, porous scales are produced, particularly under industrial conditions, and removal of carbon monoxide is no problem. However, it has been shown[3] that very careful heating can produce a non-porous scale or a scale of greatly reduced permeability to carbon monoxide. As a result, instead of showing a decarburised surface, the steel shows carbon enrichment at the surface. This clearly demonstrates that removal of the gas carbon monoxide is vital for decarburisation to occur.

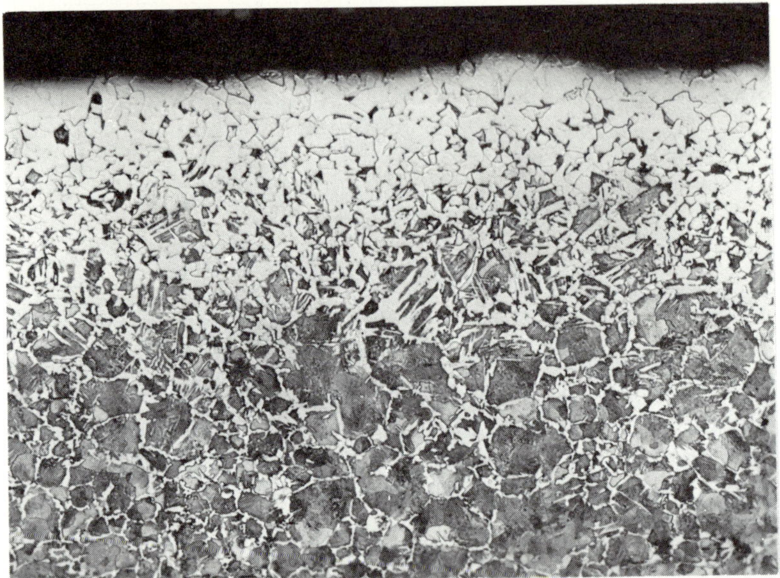

Fig. 9.1 Micrograph of 0.6% C steel after heating in air for 30 minutes at 1100 °C. ×70

Figure 9.1 shows the decarburised surface of a 0.6% C steel heated in air at 1100 °C for 30 minutes. Several points should be noted. The decarburisation occurs in the surface layers, but carbon is clearly withdrawn rapidly from the grain boundaries, and grain boundary decarburisation persists well into the specimen. Although decarburisation is a surface phenomenon, there is no clear indication of the precise position which may be taken to indicate the inner limit of the decarburised zone. This latter point is important since measurements of the 'depth of decarburisation' are used commercially to describe the condition of the steel. However, this measurement is not precise, requiring judgement of the position of the inner limit, thus introducing a human element into the measurement. Further difficulties are introduced by the effect of cooling rate and alloy content, such as manganese, on the precipitation of proeutectoid ferrite during cooling and the eventual composition of the pearlite that forms as the eutectoid structure. This is not the place to go into these factors in detail, but the above warning note should be borne in mind.

Simultaneous isothermal scaling and decarburisation

The various conditions likely to be met are shown in Fig. 9.2. Figure 9.2(a) shows the various conditions for reheating and it should be noticed that

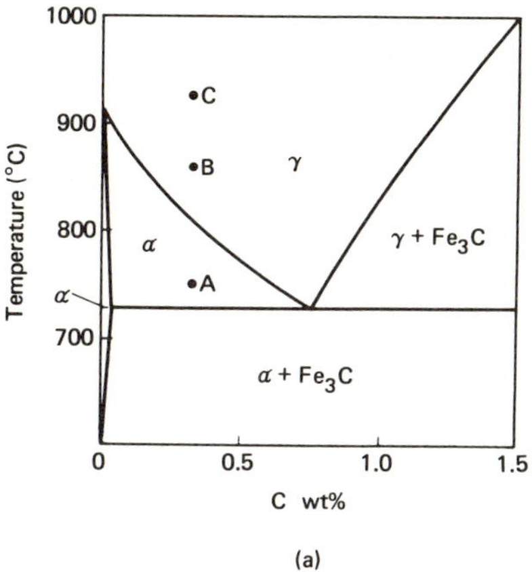

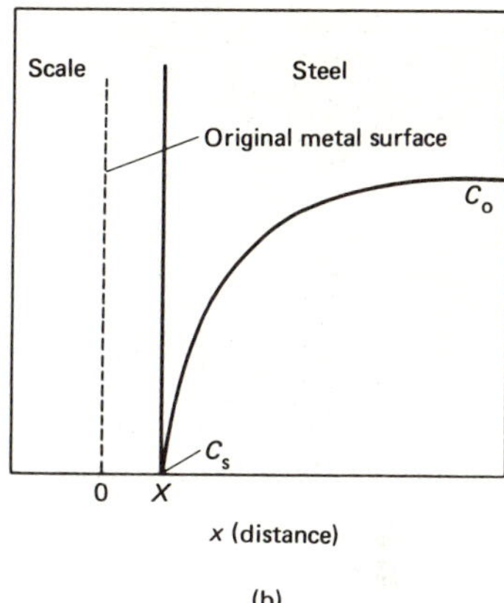

Fig. 9.2 (a) Iron–carbon phase diagram with different heating conditions A, B, and C indicated, (b) carbon profile and model for the decarburisation of plain carbon steel above 910 °C, corresponding to condition C in (a)

conditions A and B result in the formation of a surface layer of ferrite which, due to its low solubility for carbon, restricts the outward diffusion of carbon. Under these conditions, decarburisation is restricted to a very thin layer of total decarburisation at the surface. Although this may be significant in the final heat treatment of finished components, it is negligible when reheating for hot working, in which case the reheating is carried out well into the austenite range and corresponds to condition C in Fig. 9.2(a). Under these conditions, the austenitic structure is maintained, even at zero carbon content, and carbon can diffuse rapidly out of the steel forming a deep rim of decarburisation. The relevant carbon profile is shown in Fig. 9.2(b). The mechanism of decarburisation is shown in Fig. 9.3.

The temperature below which a ferrite surface layer forms on decarburisation depends on the nature and concentration of other alloying elements in the steel, but, for plain carbon steels, can be taken to be 910 °C as for pure iron. Prediction of the effect of alloying elements on this temperature is complicated by the fact that the alloying elements will undergo either denudation or concentration at the surface during scaling.

In the model shown for the carbon profile in Fig. 9.2(b), C_o is the original carbon content, C_s is the carbon content at the scale–metal interface, x is distance measured from the original metal surface, and X is the position of the scale–metal interface.

The shape of the carbon profile can be obtained by solving Fick's second law for the case of a semi-infinite slab[4].

$$\frac{\partial C}{\partial t} = D\frac{\partial^2 C}{\partial x^2} \quad \text{for } x > X \tag{9.2}$$

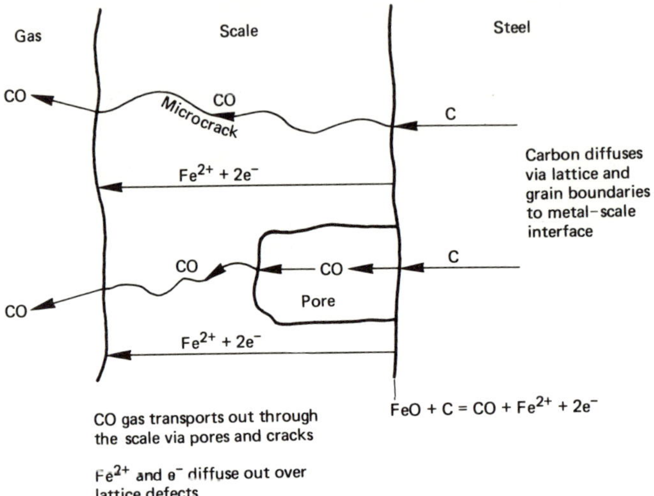

Fig. 9.3 Mechanism of decarburisation of plain carbon steel with simultaneous scaling

Assuming that the diffusion coefficient of carbon, D, is constant, the general solution of equation 9.2 for a semi-infinite slab is[4]

$$C = A + B \operatorname{erf}(x/2\sqrt{Dt}) \tag{9.3}$$

The constants A and B are eliminated by considering the boundary conditions 9.4 and 9.5 below.

$$C = C_o \quad \text{for } x > 0, t = 0 \tag{9.4}$$

i.e. the carbon concentration was initially uniform, and

$$C = C_s \quad \text{for } x = X, t > 0 \tag{9.5}$$

i.e. the carbon concentration at the metal–scale interface, C_s, is held constant in equilibrium with the scale.

Under these conditions, the solution of equation 9.3 at constant temperature is

$$\frac{C_o - C}{C_o - C_s} = \frac{\operatorname{erfc}(x/2\sqrt{Dt})}{\operatorname{erfc}(k_c/2D)^{1/2}} \tag{9.6}$$

where k_c is the 'corrosion constant' of the steel defined as

$$k_c = X^2/2t \tag{9.7}$$

in which X is the depth of metal consumed by scale formation at time t.

Equation 9.6 can be simplified by putting $C_s = 0$, since this value is extremely low ($\lesssim 0.01\%$), yielding

$$C = C_o\left[1 - \frac{\text{erfc}\,(x/2\sqrt{Dt})}{\text{erfc}\,(k_c/2D)^{1/2}}\right] \tag{9.8}$$

Using a steel of 0.85% C, 0.85% Mn, and 0.18% Si, data for k_c were obtained by direct measurement[5] and found to obey the relationship

$$k_c = 57.1 \exp(-21720/T)\ \text{mm}^2\ \text{s}^{-1} \tag{9.9}$$

In practice, the value D varies with carbon content, but values for low carbon content were found to give good agreement between calculated and measured profiles[5]. Extrapolating the data of Wells *et al.*[6] to zero carbon content yielded[5]

$$D = 24.6 \exp(-17540/T)\ \text{mm}^2\ \text{s}^{-1} \tag{9.10}$$

Combining equations 9.8, 9.9, and 9.10, carbon profiles can be calculated and, as shown in Fig. 9.4, these give excellent agreement with measured profiles.

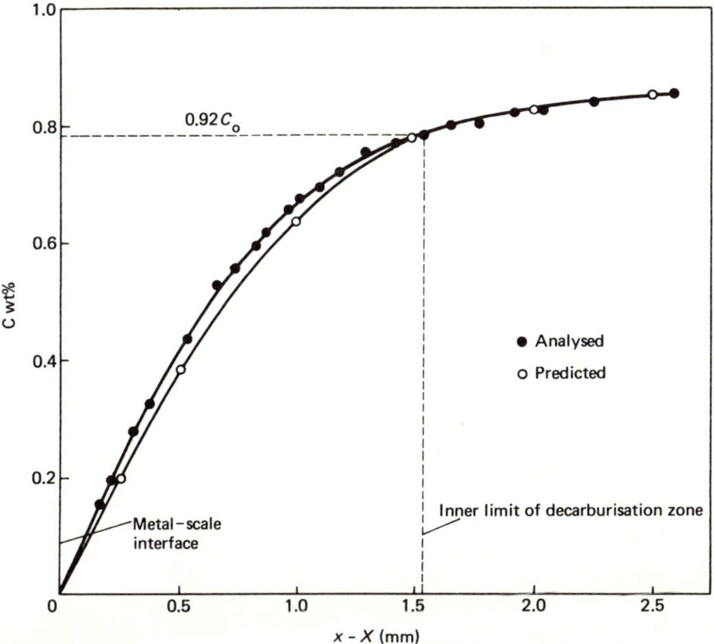

Fig. 9.4 Comparison of analysed and predicted profiles for 0.85% C steel heated at 1100 °C for 90 minutes showing the position of the inner decarburisation limit

The actual value of the depth of decarburisation is more difficult to establish since the carbon profile is smooth and it is difficult to define an inner limit. Although this could, and perhaps should, be set by defining a carbon content below which the mechanical properties of the steel are below specification, current practice relies on the choice of this position 'by eye' by an experienced metallographer. By comparing reported depths of decarburisation with the carbon profiles, the inner limit was established[5] as the position where the carbon content is 92% of the original uniform carbon content of the steel. This is also indicated in Fig. 9.4.

Denoting the position of the inner limit as $x = x^*$, where $C = C^* = 0.92C_o$, the depth of decarburisation, d, can now be established as

$$d = x^* - X = x^* - (2k_c t)^{1/2} \tag{9.11}$$

From equation 9.11, it can be seen that the value of d depends on the value of X or k_c, and in atmospheres where scaling is rapid k_c and X will be high and d will be correspondingly low. This will be discussed later.

The relationship for d, the observed depth of decarburisation, must be established for each grade of steel using parameters measured for that grade. For example, for the 0.85% C steel above, the following values are used

$$C = C^* = 0.92C_o \text{ at } x = x^* \quad \text{and} \quad C_o - C_s \approx C_o$$

Thus

$$\frac{C_o - C}{C_o - C_s} = 0.08$$

Using this value in equation 9.6, and introducing equations 9.9 and 9.10, the following relationship is obtained for x^*

$$\text{erfc}\,(x^*/2\sqrt{Dt}) = 0.08\,\text{erfc}\,[1.16\exp(-4180/T)]^{1/2} \tag{9.12}$$

Examination of equation 9.12 between 900 and 1300 °C shows that the value $x^*/2\sqrt{Dt}$ varies between 1.30 and 1.36, respectively.

In general we can put

$$x^* = 2\alpha\sqrt{Dt} \tag{9.13}$$

where α varies slightly with temperature.

This expression can now be introduced into equation 9.11 giving

$$d = [2\alpha D^{1/2} - (2k_c)^{1/2}]t^{1/2} \tag{9.14}$$

Substituting for D and k_c from equations 9.9 and 9.10 yields

$$d = 9.92\alpha\exp(-8710/T)\left[1 - \frac{10.66}{9.92\alpha}\exp(-2150/T)\right]t^{1/2} \tag{9.15}$$

By estimating the term within square brackets, and using values of α, for 900 and 1300 °C, the following expressions for d are given

$$d = 10.84\exp(-8710/T)\,t^{1/2}\ \text{mm} \quad \text{using 900 °C values}$$

and

$$d = 10.06 \exp(-8710/T)\, t^{1/2} \text{ mm} \quad \text{using } 1300\,^{\circ}\text{C values}$$

Thus, to within 4% error, the depth of decarburisation may be written, for this steel, as

$$d = 10.5 \exp(-8710/T)\, t^{1/2} \text{ mm} \tag{9.16}$$

where t is in seconds. Equation 9.16 allows the observed depth of decarburisation at constant temperature, T, between 900 and 1300 °C, to be calculated for the above 0.85% C steel. For other steels a similar equation would be valid, the difference being mainly in the pre-exponential term.

It should also be noted that, in steels of lower carbon content than the eutectoid composition, the presence of ferrite in the microstructure makes the judgement of the inner limit of decarburisation more difficult and open to greater error. Thus, the agreement between prediction and observation is expected to deteriorate at lower carbon contents.

Effect of scaling rate on decarburisation

One of the most common fallacies is to attempt to reduce decarburisation by reducing the oxygen potential in the furnace. Since the carbon content at the metal–scale interface is constant, so long as FeO remains in contact with the steel, the driving force for carbon diffusion is also constant in the presence of a scale. However, by reducing the oxygen potential of the atmosphere, the scaling rate can be reduced and this will affect the observed depth of decarburisation.

To illustrate this, calculated carbon profiles are plotted[7] in Fig. 9.5 for an 0.85% C steel heated for 90 minutes at 1050 °C for various assumed values of k_c, the corrosion constant. The profiles are plotted relative to the original metal surface and refer to values of k_c of 4.1×10^{-6} mm^2 s^{-1}, which is a realistic value for a furnace atmosphere, 4.1×10^{-5} mm^2 s^{-1}, a hypothetical value to represent a very high scaling rate, and 0 mm^2 s^{-1}, representing the case where the atmosphere just fails to form a scale but still reduces the surface carbon content to very low values.

It can be seen from Fig. 9.5 that reducing the scaling rate reduces the total metal wastage but increases the observed depth of decarburisation. This can be seen from the values of metal lost due to scaling, X, and the observed depth of decarburisation, d, given in Table 9.1.

If the product is to be machined to remove decarburisation, less metal is wasted if less aggressive atmospheres are used. Alternatively, reheating an awkwardly shaped product, such as wire, in a very aggressive atmosphere (by injecting steam) may cause the observed depth of decarburisation to be reduced. Obviously, the questions of cost and economy must carefully be considered.

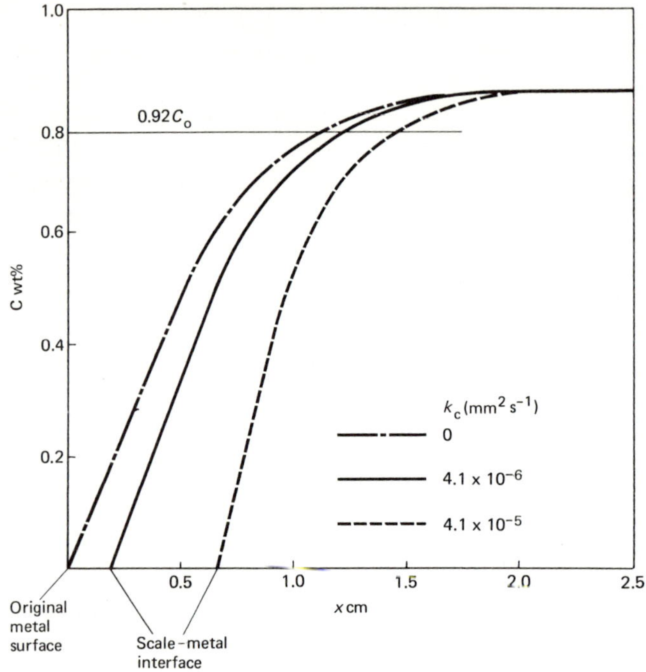

Fig. 9.5 Effect of scaling rate on depth of decarburisation, calculated for 0.85% C steel heated for 90 minutes at 1050 °C

Effect of prior decarburisation

Most steels already have a decarburised surface layer, particularly at the billet stage, before they are subjected to reheating. Equation 9.16, in fact, predicts the depth of decarburisation for the case where the steel has no initial, or prior, decarburisation. To account for such initial decarburisation, it is suggested[2,5] that equation 9.16 be rearranged as

$$d^2 = d_o^2 + 110t \exp(-17\,420/T) \text{ mm}^2 \tag{9.17}$$

Table 9.1 Calculated values for the effect of scaling rate on decarburisation for 0.85% C steel after 90 minutes at 1050 °C

Corrosion constant k_c (mm^2 s^{-1})	Depth of decarburisation (mm)	Metal removed by scaling (mm)	Total depth affected (mm)
0	1.19	0	1.19
4.1×10^{-6}	1.09	0.21	1.30
4.1×10^{-5}	0.84	0.66	1.50

where d_o is the initial depth of decarburisation at $t = 0$. Thus, from equation 9.17, if a steel with no initial decarburisation is reheated at 1200 °C for 20 minutes, the value $d^2 = 0.965$ mm^2 is given, corresponding to a depth of decarburisation of 0.98 mm. If the steel already had an initial depth of decarburisation of 0.5 mm before this treatment then, subsequent to reheating at 1200 °C for 20 minutes, the depth $d = (0.5^2 + 0.98^2)^{1/2}$ mm = 1.10 mm is expected.

Simultaneous non-isothermal scaling and decarburisation

All of the previous discussion refers to isothermal conditions. Very few commercial reheating cycles can adequately be represented by an isothermal process. Consequently, it is necessary to apply the above treatment to non-isothermal steps and to assess the decarburisation at each step[2,5].

This technique follows the way in which the effect of initial decarburisation is assessed in the previous section. In fact, the decarburisation undergone up to a certain temperature is regarded as an initial value prior to decarburisation at that temperature. Thus, if d_o is the initial decarburisation depth and d_T is the decarburisation depth at any temperature T, we have as before

$$d^2 = d_o^2 + \sum_T d_T^2 \qquad (9.18)$$

Values of d_T are assessed using equation 9.16 for the time and temperature appropriate to that step in the thermal cycle under consideration.

Conclusions

The quantitative treatment presented of the decarburisation of steels contains many assumptions, despite which it produces acceptable results in the isothermal case. To extend the treatment to the non-isothermal case, certain factors have been overlooked. For example, the existence of a temperature gradient in the steel has been ignored and both grain boundary diffusion and the variation of carbon diffusion with carbon content have not been accounted for. Nevertheless, the model can produce useful results for commercial application and a more comprehensive, computerised version should be capable of more accurate predictions.

References

1 Sachs, K. and Tuck, C. W., *ISI Publication 111*, 1968; p. 1
2 Birks, N. and Jackson, W., *JISI*, **208,** 81, 1970

3 Baud, J., Ferrier, A., Manenc, J. and Bénard, J., *Oxid. Metals*, **9,** 1, 1975
4 Crank, J., *Mathematics of Diffusion*, Oxford University Press, 1950
5 Birks, N. and Nicholson, A., *ISI Publication 123*, 1970; p. 219
6 Wells, C. *et al.*, *TAIME*, **188,** 553, 1950
7 Birks, N., *ISI Publication 133*, 1970; p. 1

Appendices

A Rigorous derivation of internal oxidation rate for planar specimens

The following is a treatment of the kinetics of internal oxidation of planar specimens which was published by Bohm and Kahlweit, *Acta Met.*, **12,** 641, 1964. This treatment is more general than that presented in the text in that it considers the counter-diffusion of solute and also allows for a finite solubility product of the internal oxide particles in the matrix. The Bohm–Kahlweit treatment is essentially a modification of an earlier analysis by Wagner, *Z. Elektrochem.*, **63,** 772, 1959, which takes into account solute diffusion but still assumes that concentrations of oxygen and solute go to zero at the reaction front.

Consider the case of atomic oxygen diffusing into the specimen from the surface ($x = 0$) in the positive x-direction and combining with the outward diffusing solute at $x = X$ to form BO particles. The two differential equations which describe this process are

$$\frac{\partial N_{\mathrm{O}}}{\partial t} = D_{\mathrm{O}} \frac{\partial^2 N_{\mathrm{O}}}{\partial x^2} \tag{A.1}$$

$$\frac{\partial N_{\mathrm{B}}}{\partial t} = D_{\mathrm{B}} \frac{\partial^2 N_{\mathrm{B}}}{\partial x^2} \tag{A.2}$$

The boundary conditions are

$$\begin{array}{lllll} t = 0 & N_{\mathrm{O}} = N_{\mathrm{O}}^{(\mathrm{S})} & \text{for } x < 0 & N_{\mathrm{B}} = 0 & \text{for } x < 0 \\ & N_{\mathrm{O}} = 0 & \text{for } x > 0 & N_{\mathrm{B}} = N_{\mathrm{B}}^{(0)} & \text{for } x \geqslant 0 \\ \text{All } t & N_{\mathrm{O}} = N_{\mathrm{O}}^{(\mathrm{S})} & \text{for } x = 0 & N_{\mathrm{B}} = N_{\mathrm{B}}^{(0)} & \text{for } x = \infty \end{array} \tag{A.3}$$

The solution of equation A.1 subject to the conditions of A.3 will be of the form*

$$N_{\mathrm{O}} = N_{\mathrm{O}}^{(\mathrm{S})} + A \,\mathrm{erf}\left(\frac{x}{2\sqrt{D_{\mathrm{O}} t}}\right) \tag{A.4}$$

where

*See J. Crank, *Mathematics of Diffusion*, Oxford University Press, 1950; p. 102–103.

$$\operatorname{erf}(Z) = \frac{2}{\sqrt{\pi}} \int_0^Z \exp(-\zeta^2)\, d\zeta \tag{A.5}$$

We assume the location of the reaction front can be written

$$X(t) = 2\gamma\sqrt{D_O t} \tag{A.6}$$

where γ is an undetermined parameter. In order to eliminate the constant A we consider the oxygen and solute concentrations at the reaction front

$$N_O(X) = N_O^m \tag{A.7}$$

$$N_B(X) = N_B^m \tag{A.8}$$

where $N_B^m N_O^m$ is the critical concentration product necessary for nucleation of the oxide precipitates. At $x = X$, equation A.4 becomes

$$N_O^m = N_O^{(S)} + A \operatorname{erf}\left(\frac{X}{2\sqrt{D_O t}}\right) \tag{A.9}$$

$$A \operatorname{erf}\left(\frac{2\gamma\sqrt{D_O t}}{2\sqrt{D_O t}}\right) = N_O^m - N_O^{(S)} \tag{A.10}$$

$$A = \frac{N_O^m - N_O^{(S)}}{\operatorname{erf}(\gamma)} \tag{A.11}$$

$$N_O(x,t) = N_O^{(S)} - \frac{N_O^{(S)} - N_O^m}{\operatorname{erf}(\gamma)} \operatorname{erf}\left(\frac{x}{2\sqrt{D_O t}}\right) \tag{A.12}$$

The form of the solution to equation A.2 subject to conditions A.3 is*

$$N_B = N_B^{(0)} + B \operatorname{erfc}\left(\frac{x}{2\sqrt{D_B t}}\right) \tag{A.13}$$

where

$$\operatorname{erfc}(Z) = \frac{2}{\sqrt{\pi}} \int_Z^\infty \exp(-\zeta^2)\, d\zeta = 1 - \operatorname{erf}(Z) \tag{A.14}$$

At $x = X$, equation A.13 becomes

$$N_B^m = N_B^{(0)} + B \operatorname{erfc}\left(\frac{2\gamma\sqrt{D_O t}}{2\sqrt{D_B t}}\right) \tag{A.15}$$

and the constant B may be written

*See Crank, p. 102–103.

$$B = \frac{N_B^m - N_B^{(0)}}{\operatorname{erfc}\left[\gamma\left(\frac{D_O}{D_B}\right)^{1/2}\right]} \tag{A.16}$$

So that equation A.13 becomes

$$N_B(x,t) = N_B^{(0)} - \frac{N_B^{(0)} - N_B^m}{\operatorname{erfc}(\theta^{1/2}\gamma)} \operatorname{erfc}\left(\frac{x}{2\sqrt{D_B t}}\right) \tag{A.17}$$

where $\theta = D_O/D_B$.

At this point the problem which remains is to evaluate the parameter γ so that equation A.6 may be used to express the subscale penetration. Bohm and Kahlweit assume the last precipitation event occurred at $x = X'$ and that the next will occur at $x = X$ and, essentially, write a materials balance for the volume element ΔX. The concentration profiles as the front moves from X' to X are assumed to be as those shown in Fig. A.1. When the front moves from X' to X the amount of oxygen diffusing into the volume element at X'

$$\left(\approx D_O \frac{\partial N_O}{\partial x}\bigg|_{X'} \Delta t\right)$$

must be equal to the amount of B precipitating from this volume element $(\approx (N_B^m - N_B')\Delta X)$ plus the amount of B which diffuses into it at X

$$\left(\approx D_B \frac{\partial N_B}{\partial x}\bigg|_{X} \Delta t\right)$$

$$-D_O \frac{\partial N_O}{\partial x}\bigg|_{X'} \Delta t = (N_B^m - N_B')\Delta X + D_B \frac{\partial N_B}{\partial x}\bigg|_{X} \Delta t \tag{A.18}$$

Differentiation of equations A.12 and A.17 yields

$$\frac{\partial N_O}{\partial x}\bigg|_{X'} = -\frac{N_O^{(S)} - N_O^m}{\operatorname{erf}(\gamma)} \exp(-\gamma^2) \frac{2}{\pi^{1/2} X'} \tag{A.19}$$

$$\frac{\partial N_B}{\partial x}\bigg|_{X} = \frac{N_B^{(0)} - N_B^m}{\operatorname{erfc}(\theta^{1/2}\gamma)} \exp(-\theta\gamma^2) \frac{2\theta^{1/2}}{\pi^{1/2} X} \tag{A.20}$$

Equations A.19 and A.20 may be substituted into equation A.18 in proceeding toward a solution of γ but the resulting equation would be unwieldy at best. However, the following approximations may be made. In general, for internal oxidation problems

$$\frac{D_B}{D_O} \ll \frac{N_O^{(S)}}{N_B^{(0)}} \ll 1 \quad \text{so that } \gamma \ll 1 \text{ and } \theta^{1/2}\gamma \gg 1$$

Under these conditions

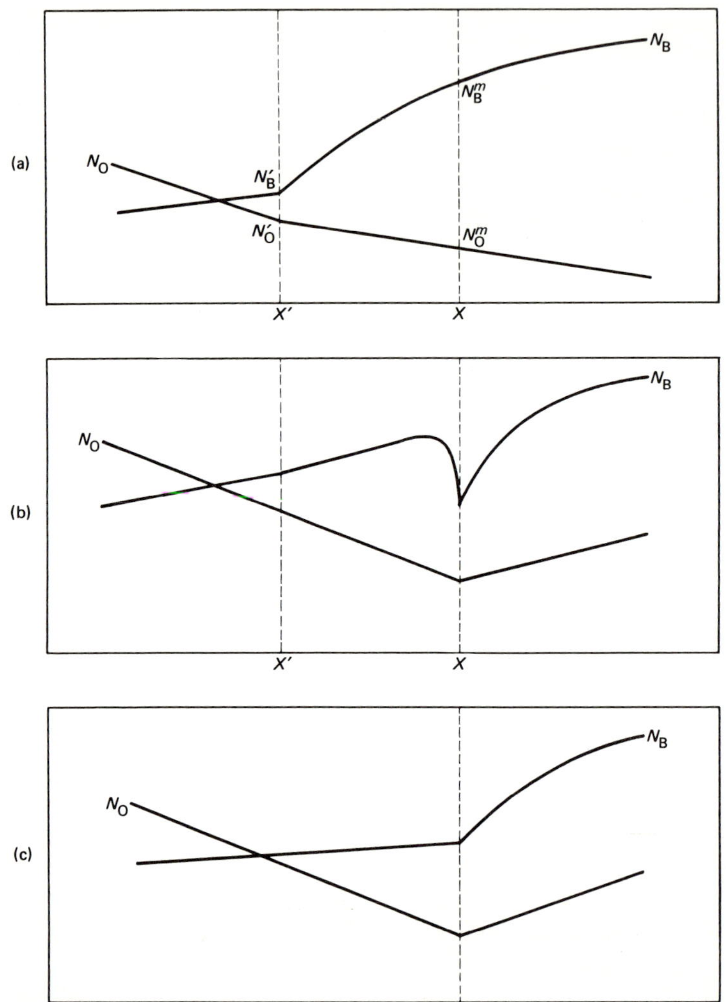

Fig. A.1 Concentration profiles for the motion of the internal oxidation front

$$\operatorname{erf}(\gamma) \approx \frac{2}{\pi^{1/2}}\gamma$$
$$\exp(\gamma^2) \approx 1 + \gamma^2 \tag{A.21}$$
$$\operatorname{erfc}(\gamma\theta^{1/2}) \approx \frac{\exp(-\gamma^2)\theta}{\frac{1}{2}\gamma\theta^{1/2}}$$

Also, for most systems, $N_O^m \ll N_O^{(S)}$. Equations A.19 and A.20 then reduce to

$$\left.\frac{\partial N_{\text{O}}}{\partial x}\right|_{X'} = -\frac{N_{\text{O}}^{(\text{S})}}{X'} \tag{A.22}$$

$$\left.\frac{\partial N_{\text{B}}}{\partial x}\right|_{X} = 2\theta\gamma^2 \frac{N_{\text{B}}^{(0)} - N_{\text{B}}^{\text{m}}}{X} \tag{A.23}$$

Equation A.18 now becomes

$$D_{\text{O}} \frac{N_{\text{O}}^{(\text{S})}}{X'} \Delta t = (N_{\text{B}}^{\text{m}} - N_{\text{B}}^{(0)})\Delta X + 2D_{\text{O}}\gamma^2 \frac{N_{\text{B}}^{(0)} - N_{\text{B}}^{\text{m}}}{X} \Delta t \tag{A.24}$$

Dividing equation A.24 by Δt and rearranging gives

$$\frac{\Delta X}{\Delta t} = \frac{1}{N_{\text{B}}^{\text{m}} - N_{\text{B}}^{(0)}} \left(D_{\text{O}} \frac{N_{\text{O}}^{(\text{S})}}{X'} - 2D_{\text{O}}\gamma^2 \frac{N_{\text{B}}^{(0)} - N_{\text{B}}^{\text{m}}}{x} \right) \tag{A.25}$$

Now writing $\Delta X/\Delta t$ using equation A.6

$$\frac{\Delta X}{\Delta t} = \frac{\text{d}X}{\text{d}t} = D_{\text{O}}^{1/2}\gamma t^{-1/2} = \frac{2D_{\text{O}}\gamma^2}{X} \tag{A.26}$$

If we now assume $X \approx X'$ in equation A.25, and compare equations A.25 and A.26, we have

$$\gamma^2 = \frac{N_{\text{O}}^{(\text{S})}}{2(N_{\text{B}}^{(0)} - N'_{\text{B}})} \tag{A.27}$$

In the above analysis it was assumed the oxide formed was BO. If we consider the oxide to be BO_ν, equation A.27 would read

$$\gamma^2 = \frac{N_{\text{O}}^{(\text{S})}}{2\nu(N_{\text{B}}^{(0)} - N'_{\text{B}})} \tag{A.28}$$

Substitution of γ from equation A.28 into equation A.6 now allows the calculation of $X(t)$. When $N'_{\text{B}} \ll N_{\text{B}}^{(0)}$ equation A.28 reduces to

$$\gamma \equiv \left(\frac{N_{\text{O}}^{(\text{S})}}{2\nu N_{\text{B}}^{(0)}} \right)^{1/2} \tag{A.29}$$

and equation A.6 gives

$$X = \left(\frac{2N_{\text{O}}^{(\text{S})} D_{\text{O}} t}{\nu N_{\text{B}}^{(0)}} \right)^{1/2} \tag{A.30}$$

which is the result obtained using the quasi-steady state approach.

Digression to consider Wagner's approach

Wagner simply equates the stoichiometric flux of oxygen and solute arriving in the region of the reaction front

$$\lim_{\varepsilon \to 0} \left(-D_O \frac{\partial N_O}{\partial x}\bigg|_{x = X - \varepsilon} = D_B \frac{\partial N_B}{\partial x}\bigg|_{x = X + \varepsilon} \right) \tag{A.31}$$

Substitution of equations A.19 and A.20 into equation A.31 with N_O^m and N_B^m being negligible relative to $N_O^{(S)}$ and $N_B^{(0)}$ gives

$$D_O \frac{N_O^{(S)}}{\text{erf}(\gamma)} \exp(-\gamma^2) \frac{2\gamma}{\pi^{1/2} X} = \nu D_B \frac{N_B^{(0)}}{\text{erfc}(\theta^{1/2}\gamma)} \exp(-\theta\gamma^2) \frac{2\theta^{1/2}\gamma}{\pi^{1/2} X} \tag{A.32}$$

which rearranges to

$$\begin{aligned} \frac{N_O^{(S)}}{N_B^{(0)}} &= \frac{\nu D_B \,\text{erf}(\gamma) \exp(-\theta\gamma^2)\, \theta^{1/2}}{D_O \,\text{erfc}(\theta^{1/2}\gamma) \exp(-\gamma^2)} \\ &= \frac{\nu \,\text{erf}(\gamma) \exp(\gamma^2)}{\theta^{1/2} \,\text{erfc}(\theta^{1/2}\gamma) \exp(\theta\gamma^2)} \end{aligned} \tag{A.33}$$

A graphical or numerical solution of equation A.33 is required to obtain γ for substitution into equation A.6 but for most real problems limiting cases may be applied.

(a) $\gamma \ll 1$ and $\gamma\theta^{1/2} \gg 1$ (i.e. the conditions of A.21 apply). For this case, equation A.33 reduces to

$$\gamma = \left[\frac{N_O^{(S)}}{2\nu N_B^{(0)}} \right]^{1/2} \tag{A.34}$$

which is equivalent to equation A.29.

(b) If $\gamma \ll 1$ and $\gamma\theta^{1/2} \ll 1$ (i.e. diffusion of alloying element is important)

$$\text{erf}(\gamma) = \frac{2}{\pi^{1/2}} \gamma$$

$$\text{erfc}(\gamma\theta^{1/2}) = 1 - \frac{2\gamma\theta^{1/2}}{\pi^{1/2}} \approx 1$$

$$\exp(\gamma^2) \approx \exp(\gamma^2\theta) \approx 1$$

Therefore, equation A.33 reduces to

$$\gamma = \frac{N_O^{(S)}}{\nu N_B^{(0)}} \frac{\sqrt{\pi}}{2} \left(\frac{D_O}{D_B} \right)^{1/2} \tag{A.35}$$

Substitution of equation A.35 into equation A.6 yields

$$X(t) = \frac{\pi^{1/2} t^{1/2} N_O^{(S)} D_O}{\nu N_B^{(0)} D_B^{1/2}} \tag{A.36}$$

B Effects of impurities on oxide defect structures

This aspect of the theory of defect structures of non-stoichiometric compounds is usually covered in the main text of books on high temperature oxidation. The subject of doping is interesting for its own sake, and it is vitally important for the study of the physical chemistry and electrochemistry of ionic compounds. In the case of an introduction to high temperature oxidation our opinion is that, since the control of oxidation rates by controlling the ionic and electronic transport properties of oxides by impurity solution is not used as a technique for the development of oxidation resistant alloys, this subject should be dealt with in an appendix which allows it to be covered adequately without over-emphasising its importance.

In the following discussion ZnO will be used as a typical n-type oxide and NiO as a typical p-type oxide.

N-type oxides

The native defect structure involving excess cations on interstitial sites with electrons in the conduction band may be represented as

$$ZnO = Zn_i^{\cdot} + e' + \tfrac{1}{2}O_2 \tag{B.1}$$

and

$$ZnO = Zn_i^{\cdot\cdot} + 2e' + \tfrac{1}{2}O_2 \tag{B.2}$$

To represent the addition of a more positive cation to the ZnO lattice consider Al_2O_3 as the impurity. The solution of Al_2O_3 in ZnO may occur in two ways.

(a) Al^{3+} ions occupy normal Zn^{2+} ion lattice sites. Since only two corresponding anion lattice sites are available for the three oxygen ions, one must be discharged releasing oxygen to the atmosphere and putting two electrons in the conduction band according to

$$Al_2O_3 = 2Al_{Zn}^{\cdot} + 2e' + 2O_O + \tfrac{1}{2}O_2 \tag{B.3}$$

(b) The introduction of extra electrons to the conduction band upsets the equilibrium between them and interstitial zinc ions according to equations B.1 and B.2. Thus Al_2O_3 also dissolves in a manner allowing some interstitial zinc ions to be eliminated according to

$$Al_2O_3 + Zn_i^{\cdot\cdot} = 2Al_{Zn}^{\cdot} + 3O_O \tag{B.4}$$

or

$$Al_2O_3 + Zn_i^{\cdot} = 2Al_{Zn}^{\cdot} + e' + 3O_O \tag{B.5}$$

The overall result of doping ZnO with Al_2O_3, i.e. the oxide of a higher valency cation, is to reduce the concentration of interstitial zinc ions and to increase the concentration of conduction band electrons, thus reducing the cationic conductivity and increasing the electronic conductivity. Such an effect would decrease the oxidation rate of an alloy on which such an oxide could form.

To represent the addition of a less positive ion onto the ZnO lattice consider the solution of lithium oxide, Li_2O. This may also occur in two ways.

(a) The two Li^+ ions occupy normal Zn^{2+} cationic sites but only one anion site is occupied. The second anion site is filled by taking $\frac{1}{2}O_2$ from the atmosphere and withdrawing electrons from the conduction band in order to ionise it.

$$Li_2O + 2e' + \tfrac{1}{2}O_2 = 2Li'_{Zn} + 2O_O \qquad (B.6)$$

(b) Since the removal of conduction band electrons upsets the equilibrium conditions of equations B.1 and B.2, an alternative mechanism would be for the two Li^+ ions to displace Zn^{2+} ions from existing cation sites and for oxygen to be evolved according to

$$Li_2O = 2Li'_{Zn} + 2Zn_i^{\cdot} + \tfrac{1}{2}O_2 \qquad (B.7)$$

This occurs together with the mechanism of equation B.6 so as to maintain the equilibrium between conduction band electrons and interstitial zinc ions.

The net result is to increase the concentration of interstitial zinc ions and reduce the concentration of conduction band electrons, thus increasing the cationic conductivity and reducing the electronic conductivity. Such doping should therefore lead to increased oxidation rates.

Similar doping reactions can also occur when the n-type conduction behaviour arises through the existence of anion vacancies such as

$$MO = V_O^{\cdot\cdot} + 2e' + \tfrac{1}{2}O_2 \qquad (B.8)$$

The corresponding doping reactions with Al_2O_3 would be

$$Al_2O_3 + V_O^{\cdot\cdot} = 2Al_M^{\cdot} + 2O_O \qquad (B.9)$$

and

$$Al_2O_3 = 2Al_M^{\cdot} + 2e' + 2O_O + \tfrac{1}{2}O_2 \qquad (B.10)$$

The doping reactions with Li_2O would be

$$Li_2O + 2e' + \tfrac{1}{2}O_2 = 2Li'_M + 2O_O \qquad (B.11)$$

and

$$Li_2O = 2Li'_M + V_O^{\cdot\cdot} + O_O \qquad (B.12)$$

P-type oxides

The intrinsic defect structure involving cation vacancies and electron holes can be expressed, using NiO as an example, as

$$\tfrac{1}{2}O_2 = O_O + V''_{Ni} + 2\dot{h} \tag{B.13}$$

The consequences of dissolving cations of higher and lower valencies than nickel may be considered similarly to the cases of n-type oxides as follows.

Dissolution of Cr_2O_3 may be achieved in two ways which occur together to preserve the equilibrium of equation B.13, i.e.

(a) Two Cr^{3+} ions occupy normal nickel sites and the extra oxygen ion from Cr_2O_3 is evolved as oxygen gas and contributes two electrons which neutralise two electron holes.

$$Cr_2O_3 + 2\dot{h} = 2Cr^{\cdot}_{Ni} + 2O_O + \tfrac{1}{2}O_2 \tag{B.14}$$

(b) The two Cr^{3+} ions occupy normal nickel ion sites and the three oxygen ions can occupy normal oxygen ion sites, thus creating a nickel ion vacancy.

$$Cr_2O_3 = 2Cr^{\cdot}_{Ni} + V''_{Ni} + 3O_O \tag{B.15}$$

Thus the dissolution of high valency cations into a cation deficient p-type semiconductor, such as NiO, leads to the creation of more cation vacancies and reduction of the concentration of electron holes, thus increasing the cation conductivity and decreasing the electronic conductivity. The net effect would be to increase the oxidation rate.

The effects of dissolving a cation of lower valency follows similar lines, e.g. using Li_2O we have

(a) The two lithium ions Li^+ occupy normal nickel ion sites, an oxygen atom is ionised from the gas phase to fill the second anion site, and two electron holes are created.

$$Li_2O + \tfrac{1}{2}O_2 = 2Li'_{Ni} + 2\dot{h} + 2O_O \tag{B.16}$$

(b) One of the two Li^+ ions occupies a nickel vacancy.

$$Li_2O + V''_{Ni} = 2Li'_{Ni} + O_O \tag{B.17}$$

The net result is that nickel vacancies are consumed and electron holes are produced to maintain equilibrium in equation B.13. Thus the cation conductivity is reduced and the electronic conductivity is increased. A corresponding reduction in oxidation rate would be expected.

It is also possible to produce p-type behaviour by having interstitial excess anions according to

$$\tfrac{1}{2}O_2 = O''_i + 2\dot{h} \tag{B.18}$$

The corresponding dissolution mechanism for Al_2O_3 and Li_2O in a p-type oxide MO with excess anions would be as follows, in order to maintain equilibrium in equation B.18.

(a) Al_2O_3 dissolution

$$Al_2O_3 = 2Al_M^{\cdot} + 2O_O + O_i'' \quad (B.19)$$

and

$$Al_2O_3 + 2\dot{h} = 2Al_M^{\cdot} + 2O_O + \tfrac{1}{2}O_2 \quad (B.20)$$

This will result in increased oxygen ion interstitials and higher ionic conductivity, together with lower concentrations of electron holes and lower electronic conductivity.

(b) Li_2O dissolution

$$Li_2O + O_i'' = 2Li_M' + 2O_O \quad (B.21)$$

and

$$Li_2O + \tfrac{1}{2}O_2 = 2Li_M' + 2\dot{h} + 2O_O \quad (B.22)$$

This will produce lower interstitial anion concentrations with correspondingly reduced ionic conductivity, together with increased electron hole concentration and associated increased electrical conductivity.

Index